SpringerBriefs in Applied Sciences and Technology

Computational Intelligence

Series Editor

Janusz Kacprzyk, Systems Research Institute, Polish Academy of Sciences,
Warsaw, Poland

SpringerBriefs in Computational Intelligence are a series of slim high-quality publications encompassing the entire spectrum of Computational Intelligence. Featuring compact volumes of 50 to 125 pages (approximately 20,000-45,000 words), Briefs are shorter than a conventional book but longer than a journal article. Thus Briefs serve as timely, concise tools for students, researchers, and professionals.

Patricia Melin · Martha Ramirez · Oscar Castillo

Clustering, Classification, and Time Series Prediction by Using Artificial Neural Networks

 Springer

Patricia Melin
Division of Graduate Studies
Tijuana Institute of Technology, TecNM
Tijuana, Baja California, Mexico

Martha Ramirez
Division of Graduate Studies
Tijuana Institute of Technology, TecNM
Tijuana, Baja California, Mexico

Oscar Castillo
Division of Graduate Studies
Tijuana Institute of Technology, TecNM
Tijuana, Baja California, Mexico

ISSN 2191-530X   ISSN 2191-5318 (electronic)
SpringerBriefs in Applied Sciences and Technology
ISSN 2625-3704   ISSN 2625-3712 (electronic)
SpringerBriefs in Computational Intelligence
ISBN 978-3-031-71100-8   ISBN 978-3-031-71101-5 (eBook)
https://doi.org/10.1007/978-3-031-71101-5

Preface

In this book, a new model for the clustering, classification, and time series prediction by using artificial neural networks to computationally simulate the behavior of the cognitive functions of the brain is presented. This model focuses on the study of intelligent hybrid neural systems and their use in time series analysis and decision support systems.

In recent years, attention has been paid to the factors that directly impact well-being and how to evaluate it through the use of quantitative and qualitative indicators that consider the management of uncertainty or multivariate risk factors, which if materialized they could cause the impossibility of guaranteeing the compliance level with organizational goals and objectives.

Which is why, for their administration, they are frequently divided into multiple categories according to their impact and consequences. On the other hand, global indicators are dynamic, and sometimes, their correlation is uncertain because they depend largely on a combination of economic, social, and environmental factors mainly.

Therefore, through the development of eight case studies, multiple time series related to the following problems are analyzed: traffic accidents, air quality, and multiple global indicators (energy consumption, birth rate, mortality rate, population growth, inflation, unemployment, sustainable development, and quality of life).

The main contribution consists of a Generalized Type-2 fuzzy integration of multiple indicators (time series) using both supervised and unsupervised neural networks and a set of Type-1, Interval Type-2, and Generalized Type-2 fuzzy systems.

The results obtained show the advantages of using the proposed model to Generalized Type-2 fuzzy integration of multiple time series attributes, enabling the end user to consider during the decision-making process different aspects of future planning and long-term governance strategies, talent management, industry strength, anti-corruption, and political structure, to take advantage of the latest opportunities presented by globalization.

This research work was partially funded by CONAHCYT and TecNM-Tijuana Institute of Technology, and we would like to express our gratitude to both institutions. In addition, we would like to thank Prof. Janusz Kacprzyk for always supporting and encouraging us to perform good research in the computational intelligence area.

Tijuana, Mexico   Prof. Patricia Melin
April 2024   Dr. Martha Ramirez
                                                                     Prof. Oscar Castillo

Contents

Chapter 1
Introduction to Prediction with Neural Networks

The human being can solve multiple problems simultaneously using different parts of the brain, which by itself represents a complex subject of study. For several decades, a great challenge has attracted the attention of the scientific community, causing different areas of study to join forces to understand how the human brain works.

Among the advances recently promoted in medicine is brain plasticity [1, 2], which adds to the work done on cognitive flexibility, which could be compared to the ability of an individual to adapt to changes.

Leading to the question of how prepared an individual is to solve a previously known problem but with new input data or how that same individual would face the resolution of new problems.

The volume of data generated in the world increases exponentially, however the level of access to basic digital services frequently increases moderately.

Meanwhile information has been considered a valuable resource for decades, a sense of immediacy currently prevails, it is necessary to have the greatest number of indicators to be able to make decisions that consider multiple factors seeking to obtain the best possible result.

In addition to the above, the use of tools for the analysis of historical data is also entering different fields of public, private and social sector industries [3, 4], which although they have personnel to carry out the operational activities generated to comply with the institutional goals and objectives [5].

Then, these are achieved at the end of a determining cycle and the corresponding performance indicators are recorded, in most cases, these data are exclusively compared with indicators of the previous period.

Since taking the observed data and organizing them sequentially over time in periods (time series) results in an activity complementary, and since it is not a priority attention, it generally stagnates in the queue of pending matters to be addressed, in other words, sometimes the urgent is weighted first over the important.

P. Melin et al., *Clustering, Classification, and Time Series Prediction by Using Artificial Neural Networks*, SpringerBriefs in Computational Intelligence, https://doi.org/10.1007/978-3-031-71101-5_1

Mostly systems are dynamic and involve a constant updating of all components, including people's knowledge, so starting from the fact that sometimes it is necessary to relearn concepts that were assumed to be mastered and combine them with experience acquired over the years, there is resistance to change.

It generates discomfort and it is not easy, it leads you to question, doubt and sometimes dissatisfaction.

However, all these beliefs are thoughts that live in the mind, which leads us to wonder what happens when a person's mind is busy, or they are not aware of what is happening around them and they must make a decision.

In other words, how can you identify which factors will determine the inference process to decide: temperamental strength, experience, or emotions.

An improvement to this decision-making process could be the integration of a technological tool that provides a decision proposal by simulating the cognitive processes of a person when deciding.

Thus, being able to compare whether the decision proposed by the person is congruent with the proposal of the intelligent system and promote an exhaustive analysis prior to the decision, which at all times will be carried out by the person.

When talking about decision-making, many scientific researchers, academic staff, or industry professionals emphasize the arduous path they go through to gather information in a timely manner and purify relevant and significant data.

Once this process of analysis and selection ends, they must face the challenge of considering the greatest number of criteria that guarantee compliance with the objectives or goals set, seeking that this decision-making meets the aspects of effectiveness and efficiency.

In addition to being transparent and sustainable, for the sake of promote continuous improvement.

Furthermore, to the fact that the degree of ignorance about a current or historical event is a determining factor when a person decides since the future conditions of the environment where a specific process develops will depend on it.

An aspect that increases the level of uncertainty is the human factor, since a decision is influenced by the level of mental, emotional, or physical uncertainty.

Additionally, to the considerations that imply that the decision maker was unaware of the information as mentioned above, there is also the case that the information is incorrect, incomplete, or obsolete, which leads to there being no clarity or confidence in the relationship of a fact-consequence, which leads to an unknown final result.

Typically, in several areas of knowledge, researchers converge in expressing in words from their area of experience how doubt, indecision, concern, or insecurity directly influences decision-making, encompassing these aspects in a lack of certainty or uncertainty.

Also, when there is a true value of a result between an interval of values, it could be considered uncertainty.

In this case, uncertainty can be applied to the prediction of future events or values, also in measurements and when the information is unknown. At this point, it is worth noting that uncertainty occurs in the registration, storage, data analysis and extraction of characteristics or selection of attributes, regardless of the manual or automatic method used.

The main reason to consider managing uncertainty is to make better decisions, predict what will happen in the future and look for answers to questions directly in the data.

For these activities, multidisciplinary methods are available, such as intelligent hybrid techniques through which the future values of a sequence of data can be estimated (prediction).

Besides, the search to find elements that meet certain characteristics based on their attributes (classification) and it is also possible to discover elements that share similar or similar properties (clustering).

A determining factor to evaluate the advantages of applying a computational model is that it is user-friendly and that its use is simple, always seeking to promote well-being in people by improving their quality of life; and be beneficial for the business cycle that is being implemented.

Thus, innovation is considered in the diversity of computational models applied to find solutions adapted to anyone's problems.

In turn, decision-making itself represents a complex problem for organizations. Primarily, when a person considers the methodology for the fulfillment of objectives and sees the need to implement a line of action, which would have repercussions at a strategic level and could cause a significant impact on their environment.

In general, there is also the limitation that the person who is at the highest level of the chain of command, has the responsibility of processing the information generated by different sources simultaneously, and must also document the analysis process carried out to make the final decision.

As economic and business conditions vary over time, decision makers must find ways to stay informed about the effects such changes will have on their operations. One technique that decision makers can use to assist in planning the level of future operational needs is prediction.

While numerous prediction methods have been designed, they all have a common objective, to make predictions of future events, so that these projections can then be incorporated into the decision-making process.

Thus, looking for how these cognitive functions influence the decision-making process, has promoted the development and research applied to the search and creation of new bioinspired computational models based on these advances in medicine. An example of this are artificial neural networks (ANNs), which have been maintained for several decades as a robust support for decision-making [6–8].

If we focus exclusively on the problem of analyzing information for decision-making, it is feasible to consider using ANNs among the wide range of intelligent computing techniques.

Since they were created taking as inspiration the functioning of the human brain, being this last factor, which makes them the ideal method to simulate the interaction and flow of information that a person normally performs in the moment prior to decision-making.

Although, as part of the decision-making process we must deal with changes in the variables or attributes of the information, thus it is also necessary to consider the factor of ambiguity or implicit uncertainty present at the time of the decision, because this factor directly affects the desired results [9–11].

In addition, the constant search for methods to solve complex problems has encouraged the use of models that use Fuzzy Inference Systems (FISs) for more than 50 years, so that they operate in a computer system and manage the uncertainty factor [12–14].

Thus, it is a human need to have technological tools that help accelerate the analysis times of historical data, better known as time series.

Though the interpretation of the results produced by these platforms must necessarily be interpreted and endorsed by an expert in the function, does not leave aside the fact that one of the benefits of these systems inspired by human behavior is to act as a tool to carry out the resolution of complex problems.

Currently there are multiple techniques that consider aspects of intelligent computing to perform clustering [15–17].

Clustering is considered a technique that helps find segments or groups of similar information in a data set or identify hidden patterns [18]. Likewise, computational models that rely on a time series to predict the next values on a timeline have also proliferated.

For both cases, Artificial Neural Networks (ANNs) are frequently used, which has been extensively investigated because one of their main characteristics is that they learn based on non-linear relationships between the inputs they have and the desired outputs, in addition to their great capacity to pattern recognition [19–21].

It is natural when facing the attention of variables in a real environment, that it is required to integrate several techniques to try to solve complex problems [22, 23].

The question arises as to what would happen if several models of these computational intelligent techniques are combined, to integrate a model that can simulate cognitive functioning of the human brain.

In which each of these ANNs represents a process implicit in the mental routine that each of us performs when thinking and the fuzzy systems simulate the decision-making.

An important factor that motivates us to carry out this research is the constant need to have a computational tool based on a new bioinspired model that simulates the cognitive functions that are activated while a person analyzes the information in their environment prior to deciding.

In addition to consider the challenges that appear when integrating information from different sources and consider multiple variables. This means that by using different intelligent hybrid methods we try to simulate the way in which a person groups and classifies data and infers what a future value could be for a given variable.

Furthermore, another fundamental aspect is the person's experience, over time they learn and master the operation of certain processes, which are constantly changing, so it may happen that a response that previously worked now no longer works, from the simple fact of perceiving a situation.

So, if the information systems are not aligned with the growth of the data being managed, it can be an additional factor that works against efficient decision-making.

As it is a partial simulation of the cognitive functioning of the human brain, it is noteworthy that human interaction when making decisions is an irreplaceable variable by its very nature. It is time to put aside technical practicality and focus on the collective human experience and emphasize that the obtained results.

However significant and invaluable they may be, do not replace the contribution of the human expert, regardless of the area of knowledge in question.

We can highlight that the main contribution of our research is the combination of ANN models and multiple FISs, which operate collaborating with each other in a nested or hierarchical manner, as well as individually.

It is intended to simulate to a certain extent the behavior of cognitive functions while making decisions based on the clustering, classification, and prediction of variables of a data set (time series), by using competitive neural networks and Self-organizing maps (SOM) to generate groups or classes based on the similarities of historical data [24].

Also, to estimate future values incorporate the use of Nonlinear autoregressive neural networks with exogenous (NARX) [25, 26] and Nonlinear autoregressive (NAR) [27], in addition to integrating the results and classifying variables using Type-1 and Type-2 fuzzy systems [28–30].

This approach differs from most existing intelligent computational methods [31–33], in that to carry out the clustering, classification and time series prediction, it focuses on combining supervised and unsupervised algorithms to carry out the training of the neural networks.

Additionally, the model is complemented with the use of fuzzy systems to carry out the classification and integration of the obtained results.

Which represents an innovation compared to the design of most computational models in the literature that only use supervised training algorithms to perform time series prediction and on the other hand use unsupervised training algorithms to classify data.

In addition to the above, one of the advantages of the proposed model that is of main relevance today is that this model considers a multidisciplinary approach for the management of uncertainty in decision-making, which represents a great challenge for people, organizations, or governments.

Besides, in the aspect of the business or operation cycle, it contributes to the integration of results in an effective and efficient manner.

This book is formed in the following fashion: In Chap. 2, we present a brief compendium of theoretical concepts related to this research. A concise description of the problem is shown in Chap. 3. The method is outlined in Chap. 4. Experiments and a brief discussion of results are indicated in Chaps. 5 and 6. Lastly, Chap. 7 outlines the conclusions.

References

1. Bonfanti, L., Charvet, C.J.: Brain plasticity in humans and model systems: advances, challenges, and future directions. Int. J. Mol. Sci. **22**(17), 9358 (2021). https://doi.org/10.3390/ijms22 179358
2. Memmi, D.: Connectionism and artificial intelligence as cognitive models. AI Soc. **4**, 115–136 (1990). https://doi.org/10.1007/BF01889639
3. Antonio Morente-Molinera, J., Wang, Y., Gong, Z.-W., Morfeq, A., Al-Hmouz, R., Herrera-Viedma, E.: Reducing criteria in multicriteria group decision-making methods using hierarchical clustering methods and fuzzy ontologies. IEEE Trans. Fuzzy Syst. **30**(6), 1585–1598 (2022). https://doi.org/10.1109/TFUZZ.2021.3062145
4. Sharma, J., Arora, M., Sonia, S., Alsharef, A.: An illustrative study on multi criteria decision making approach: analytical hierarchy process. In: Proceedings of the in 2022 2nd International Conference on Advance Computing and Innovative Technologies in Engineering, ICACITE 2022, pp. 2000–2005. Institute of Electrical and Electronics Engineers Inc. (2022). https://doi.org/10.1109/ICACITE53722.2022.9823864
5. Thakkar, J.J.: Introduction. In: Multi-Criteria Decision Making. Studies in Systems, Decision and Control, Vol. 336, pp. 1–25. Springer, Singapore (2021). https://doi.org/10.1007/978-981-33-4745-8_1
6. Blanchet, O., Ramirez, M., Gutierrez, M., Quintero, J., Mancilla, A., Melin, P.: A hybrid approach with modular neural networks and fuzzy logic for time series prediction. In: Proceedings of the 2006 International Conference on Artificial Intelligence, ICAI'06, pp. 591–597 (2006)
7. Siłka, J., Wieczorek, M., Woźniak, M.: Recurrent neural network model for high-speed train vibration prediction from time series. Neural Comput. Appl. **34**(16), 13305–13318 (2022). https://doi.org/10.1007/s00521-022-06949-4
8. Sohrabi Geshnigani, F., Golabi, M.R., Mirabbasi, R., Tahroudi, M.N.: Daily solar radiation estimation in Belleville station, Illinois, using ensemble artificial intelligence approaches. Eng. Appl. Artif. Intell. **120**, 105839 (2023)
9. Rahman, M.M., et al.: A comprehensive study and performance analysis of deep neural network-based approaches in wind time-series forecasting. J. Reliab. Intell. Environ. **9**(2), 183–200 (2023). https://doi.org/10.1007/s40860-021-00166-x
10. Hu, Y., Sun, X., Nie, X., Li, Y., Liu, L.: An enhanced LSTM for trend following of time series. IEEE Access **7**, 34020–34030 (2019). https://doi.org/10.1109/ACCESS.2019.2896621
11. Sehrawat, P.K., Vishwakarma, D.K.: Comparative analysis of time series models on COVID-19 predictions. In: International Conference on Sustainable Computing and Data Communication Systems, ICSCDS 2022: Proceedings, pp. 710–715. Institute of Electrical and Electronics Engineers Inc. (2022). https://doi.org/10.1109/ICSCDS53736.2022.9760992
12. Chimatapu, R., Hagras, H., Kern, M., Owusu, G.: Hybrid deep learning type-2 fuzzy logic systems for explainable AI. In: Proceedings of the 2020 IEEE International Conference on Fuzzy Systems (FUZZ-IEEE), Glasgow, UK, pp. 1–6 (2020). https://doi.org/10.1109/FUZZ48 607.2020.9177817
13. Miramontes, I., Melin, P.: Interval type-2 fuzzy approach for dynamic parameter adaptation in the bird swarm algorithm for the optimization of fuzzy medical classifier. Axioms **11**(9), 485 (2022). https://doi.org/10.3390/axioms11090485
14. Sobhi, S., Dick, S.: An investigation of complex fuzzy sets for large-scale learning. Fuzzy Sets Syst. **471**, 108660 (2023). https://doi.org/10.1016/j.fss.2023.108660
15. Cherif, A., Cardot, H., Boné, R.: SOM time series clustering and prediction with recurrent neural networks. Neurocomputing **74**(11), 1936–1944 (2011). https://doi.org/10.1016/j.neu com.2010.11.026
16. Sfetsos, A., Siriopoulos, C.: Combinatorial time series forecasting based on clustering algorithms and neural networks. Neural Comput. Appl. **13**(1), 56–64 (2004). https://doi.org/10.1007/s00521-003-0391-y

17. Xu, C., Huang, H., Yoo, S.: A deep neural network for multivariate time series clustering with result interpretation. In: Proceedings of the International Joint Conference on Neural Networks, pp. 1–8. Institute of Electrical and Electronics Engineers Inc. (2021). https://doi.org/10.1109/IJCNN52387.2021.9533427

18. Jang, J.: Fuzzy inference systems. In: Neuro-Fuzzy and Soft Computing: A Computational Approach to Learning and Machine Intelligence, pp. 73–90. Prentice Hall, Upper Saddle River (1997)

19. Prakhar, K., Sountharrajan, S., Suganya, E., Karthiga, M., Sathis Kumar, B.: Effective stock price prediction using time series forecasting. In: Proceedings of the 2022 6th International Conference on Trends in Electronics and Informatics, ICOEI 2022: Proceedings, pp. 1636–1640. Institute of Electrical and Electronics Engineers Inc. (2022). https://doi.org/10.1109/ICOEI53556.2022.9776830

20. Moghar, A., Hamiche, M.: Stock market prediction using LSTM recurrent neural network. In: Procedia Computer Science, pp. 1168–1173. Elsevier (2020). https://doi.org/10.1016/j.procs.2020.03.049

21. Wei, D.: Prediction of stock price based on LSTM neural network. In: Proceedings of the 2019 International Conference on Artificial Intelligence and Advanced Manufacturing, AIAM 2019, pp. 544–547. Institute of Electrical and Electronics Engineers Inc. (2019). https://doi.org/10.1109/AIAM48774.2019.00113

22. Egrioglu, E., Bas, E.: A new hybrid recurrent artificial neural network for time series forecasting. Neural Comput. Appl. 35(3), 2855–2865 (2023). https://doi.org/10.1007/s00521-022-07753-w

23. Li, T., Hua, M., Wu, X.: A hybrid CNN-LSTM model for forecasting particulate matter (PM2.5). IEEE Access 8, 26933–26940 (2020). https://doi.org/10.1109/ACCESS.2020.2971348

24. Yan, J., Zhang, C., Li, Y.: A clustering method for power time series curves based on improved self-organizing mapping algorithm. In: Proceedings of the 2023 IEEE 3rd International Conference on Electronic Technology, Communication and Information, ICETCI 2023, pp. 451–455. Institute of Electrical and Electronics Engineers Inc. (2023). https://doi.org/10.1109/ICETCI57876.2023.10176414

25. Yang, Y., Solomin, E., Zhou, Y.: Non-linear autoregressive neural network based wind direction prediction for the wind turbine Yaw system. In: Proceedings of the 2023 International Conference on Industrial Engineering, Applications and Manufacturing, ICIEAM 2023, pp. 119–123. Institute of Electrical and Electronics Engineers Inc. (2023). https://doi.org/10.1109/ICIEAM57311.2023.10138978

26. Carreon-Ortiz, H., Valdez, F., Melin, P., Castillo, O.: Architecture optimization of a non-linear autoregressive neural networks for mackey-glass time series prediction using discrete mycorrhiza optimization algorithm. Micromachines 14(1), 149 (2023). https://doi.org/10.3390/mi14010149

27. Melin, P., Monica, J.C., Sanchez, D., Castillo, O.: A new prediction approach of the COVID-19 virus pandemic behavior with a hybrid ensemble modular nonlinear autoregressive neural network. In: Soft Computing, vol. 27, pp. 2685–2694. Springer, New York (2023). https://doi.org/10.1007/s00500-020-05452-z

28. Melin, P.: Introduction to type-2 fuzzy logic in neural pattern recognition systems. In: Modular Neural Networks and Type-2 Fuzzy Systems for Pattern Recognition. Studies in Computational Intelligence, Vol. 389, pp. 3–6. Springer, Berlin (2012). https://doi.org/10.1007/978-3-642-241 39-0_1

29. Zhang, Z.: Trapezoidal interval type-2 fuzzy aggregation operators and their application to multiple attribute group decision making. Neural Comput. Appl. 29(4), 1039–1054 (2018). https://doi.org/10.1007/s00521-016-2488-0

30. Chen, Z., Wan, S., Dong, J.: An efficiency-based interval type-2 fuzzy multi-criteria group decision making for makeshift hospital selection. Appl. Soft Comput. 115, 108243 (2022). https://doi.org/10.1016/j.asoc.2021.108243

31. Xu, S., Li, W., Zhu, Y., Xu, A.: A novel hybrid model for six main pollutant concentrations forecasting based on improved LSTM neural networks. Sci. Rep. 12(1), 1–17 (2022). https://doi.org/10.1038/s41598-022-17754-3

32. Ding, X., Hao, K., Cai, X., Tang, X.S., Chen, L., Zhang, H.: A novel similarity measurement and clustering framework for time series based on convolution neural networks. IEEE Access **8**, 173158–173168 (2020). https://doi.org/10.1109/ACCESS.2020.3025048
33. Mónica, J.C., Melin, P., Sánchez, D.: Genetic optimization of ensemble neural network architectures for prediction of COVID-19 confirmed and death cases. In: Studies in Computational Intelligence, Vol. 940, pp. 85–98. Springer, New York (2021). https://doi.org/10.1007/978-3-030-68776-2_5

Chapter 2
Literature Review on Prediction with Neural Networks

This chapter presents a theoretical summary of the methods used to design the proposed model, focusing on the general concepts of neural networks and fuzzy systems to address intelligent computing techniques as bioinspired methods. In addition, the topics of time series, cognitive flexibility, and decision are presented.

2.1 Time Series

When we refer to the data and how it is organized by period, type, importance, among other characteristics, it is difficult to find an area of knowledge where this type of information is not considered relevant. Therefore, often when we talk about data that is arranged in a chronological sequence, in a general sense we refer to it as a time series.

Therefore, a time series is considered as an option to represent data in a structured way, to visualize the changes recorded over time and identify its trend, seasonality or simply observe the shape it presents.

In addition, time series are generally used to analyze and predict the behavior of a given variable, since depending on the model selected and the nature of the data, the quality of the prediction varies, which is why it is a current challenge in the research area [1, 2].

The importance of time series analysis lies in answering questions through the data. The objectives of this type of analysis are diverse, highlighting prediction, process control, process simulation, and the generation of new physical or biological theories.

© The Author(s), under exclusive license to Springer Nature Switzerland AG 2024
P. Melin et al., *Clustering, Classification, and Time Series Prediction by Using Artificial Neural Networks*, SpringerBriefs in Computational Intelligence,
https://doi.org/10.1007/978-3-031-71101-5_2

So, in the case of prediction, we seek to have complete certainty in the next value of a variable based on the past values, when this is not possible, we are facing a non-deterministic or random time series.

Also, prediction is widely used in the field of engineering and economics, including public health and health surveillance in the latter branch. Another field in which prediction using time series is applied is meteorology or the prediction of other natural phenomena.

2.2 Neural Networks

Artificial Neural Networks (ANNs) emerged in the 1940s, in the same way as computational theories, cybernetic applications, efforts to carry out cognitive simulations and the study of computer science [3–5].

Among the computational models that simulate the way the brain operates, the most used are ANNs that are trained to recognize patterns while learning from data [6, 7].

For the case where the algorithm uses input and output data as an example to carry out training, we can point out that it is a supervised neural network model [8–10], unlike unsupervised neural network models [11], where only input data are used to form clusters that represent characteristics of the data through classes [12, 13].

The prediction of the behavior of time series using ANNs has been extensively investigated because they learn based on non-linear relationships between the inputs available and the desired outputs and their great capacity for pattern recognition.

Although ANNs are a powerful tool for processing infinity of data and their success has been demonstrated in applications from different areas of knowledge [14–16]. Today, we can find different applications that consider ANNs to solve different problems and have proven to be effective and precise [17–19].

2.3 Type-2 Fuzzy Systems

In general terms about the theory of fuzzy logic, we can point out that its basic definitions apply to Type-1 and Type-2 fuzzy sets. A Type-2 fuzzy system is integrated with fuzzy if–then rules and membership functions where the antecedent or consequent has Type-2 fuzzy sets that are composed of Type-1 fuzzy sets.

It is a generalization of Type-1 fuzzy logic since in addition to considering the uncertainty in the linguistic variables, this uncertainty is also considered in the definition of the membership functions [20].

A General Type-2 fuzzy set A is formed by a primary variable x with domain X and a secondary variable u with domain J_x. It can be mathematically expressed as (2.1):

$$A = \{((x, u), u_A(x, u))|x \in X, u \in J_{x,} J_{x,} \subseteq [0,1]\} \tag{2.1}$$

The Footprint of Uncertainty (FOU) is mathematically expressed as (2.2):

$$FOU(A) = \left\{(x, u)|x \in X \text{ and } u \in \left[\underline{\mu}_{A}(x), \overline{\mu}_A(x)\right]\right\} \tag{2.2}$$

where the $\underline{\mu}_{A}(x)$ and $\overline{\mu}_A(x)$ are the lower and upper membership functions, respectively.

Now, in an Interval Type-2 Mamdani FIS a process similar to that of a Type-1 is carried out, the main difference is the activation forces of the upper and lower rules, as mathematically expressed in (2.3):

$$R^l : \text{IF } x_1 \text{ is } F_1^l \text{ and} \ldots \text{and } x_p \text{ is } F_p^l \text{ THEN } y \text{ is } G^l \tag{2.3}$$

where $l = 1, \ldots, M$ [21].

A fuzzy system is composed of a knowledge base represented by fuzzy rules, a database that stores the parameters and specifications of the membership functions, and a mechanism that simulates reasoning.

It means, a procedure to perform inference when we apply a fuzzy rule on the input values and calculate a result, which is usually a fuzzy set. Therefore, a method is needed to use defuzzification to extract a crisp value.

In other words, it refers to obtaining a weighted average of the output results of the Type-1 FIS that are embedded in the Type-2 FIS, where the weights correspond to the memberships in the reduced type set.

Thus, Type-2 fuzzy logic is an approach that allows representing the uncertainty in an information system because it offers to manage uncertainty in the linguistic variables by modeling vagueness and combating the lack of reliability of the information.

Furthermore, an interval represents the degree of membership of the function (it consists of two limits between 0 and 1) [22–24].

2.4 Multi-Criteria Decision Making

Multiple-criteria decision-making, or multiple-criteria decision belongs to the area of operations research (OR). As drivers of the consolidation of successful projects, people capable of making multi-criteria decisions are required.

Which help to address the greatest number of possibilities and the use of existing resources, that is, people capable of offering consistent and efficient responses to a given problem.

Their purpose is to be a growing element in quantitative analytical activity for the resolution of complex problems through a systematic and formal evaluation.

Among the variables to be evaluated are economic aspects and quality measurement, considering the optimization of resources, the experience of experts, group decisions and the application of computing techniques. The above aims to make coherent decisions in situations of uncertainty [25–27].

Now, when the situation arises in which a system must decide, the approach that is frequently used particularly in artificial intelligence, is to use basic connectors that operate from a general perspective to simulate decisions in a computer system.

It is also common for the decision problem to have several criteria. Between the aspects to consider simulating the environment through an information system, it is necessary to analyze the environment and consider whether the information sources are reliable.

For the case in which there is only one source of information (a sensor or an expert) their reliability is often insufficient. On the contrary, if there are numerous sources (sensors or experts), then it is necessary to combine them to improve the reliability and accuracy of the data [28, 29].

When an indicator (criterion) is used to make a prediction, its usefulness can be established if the obtained result is better when compared against the result of a prediction made ignoring this indicator.

There are other types of indicators known as early warning or vulnerability indicators, which are useful for policy makers when faced with situations of serious recessions, international events through which it is possible to evaluate the vulnerabilities of a country. In a global economy, countries' vulnerabilities accumulate and potentially spread between them [30].

Commonly to make a ranking by which the strengths and weaknesses of each nation are compared, it is necessary to select an appropriate classification mechanism that allows for a deep and accurate analysis, which also considers multiple factors to evaluate the development of a country [31–33].

2.5   Cognitive Flexibility

An ability that occurs when alternative responses are generated through changing a thought, choosing to use appropriate information, and understanding situations when deciding, is known as cognitive flexibility [34, 35].

One theory is that cognitive flexibility arises from the coherent interaction of processes central to an executive function, such as higher-order processes, including attention, working memory, inhibition, and switching.

In other words, it is understood as the ability to effectively adapting a set of cognitive and behavioral strategies in response to changes in the task or modification of environmental demands.

Cognitive flexibility is gradual, in the case in which individuals show rigid cognitions and perseverative behaviors often called cognitive inflexibility or 'rigidity', they could be said to have less cognitive flexibility.

Conversely, on the other hand, individuals can modify routines in their behavior and mental guides to categorize ideas and prioritize task demands based on multiple concepts, in addition to establishing non-obvious relationships instead of copying them in the form in which they were originally learned have greater cognitive flexibility [36].

References

1. Gupta, K., Tayal, D.K., Jain, A.: An experimental analysis of state-of-the-art time series prediction models. In: Proceedings of the 2022 2nd International Conference on Advance Computing and Innovative Technologies in Engineering, ICACITE 2022, pp. 44–47. Institute of Electrical and Electronics Engineers Inc. (2022). https://doi.org/10.1109/ICACITE53722.2022.9823455

2. Pirani, M., Thakkar, P., Jivrani, P., Bohara, M.H., Garg, D.: A comparative analysis of ARIMA, GRU, LSTM and BiLSTM on financial time series forecasting. In: IEEE International Conference on Distributed Computing and Electrical Circuits and Electronics, ICDCECE 2022, pp. 1–6. Institute of Electrical and Electronics Engineers Inc. (2022). https://doi.org/10.1109/ICDCECE53908.2022.9793213

3. Castillo, O., Melin, P.: Hybrid intelligent systems for time series prediction using neural networks, fuzzy logic, and fractal theory. IEEE Trans. Neural Netw. 13(6), 1395–1408 (2002). https://doi.org/10.1109/TNN.2002.804316

4. Ovchynnikova, O., Belovsky, C., Khan, O.: Neural network forecasting of international population migration. In: Proceedings of the 2021 11th International Conference on Advanced Computer Information Technologies, ACIT 2021—Proceedings, pp. 147–152. Institute of Electrical and Electronics Engineers Inc. (2021). https://doi.org/10.1109/ACIT52158.2021.9548420

5. Kan, V., Alsova, O.: Forecasting meteorological indicators based on neural networks. In: Proceedings of the 2022 IEEE International Multi-Conference on Engineering, Computer and Information Sciences, SIBIRCON 2022, pp. 1620–1625. Institute of Electrical and Electronics Engineers Inc. (2022). https://doi.org/10.1109/SIBIRCON56155.2022.10017124

6. Valdez, F., Melin, P., Castillo, O.: Modular neural networks architecture optimization with a new nature inspired method using a fuzzy combination of particle swarm optimization and genetic algorithms. Inf. Sci. 270, 143–153 (2014). https://doi.org/10.1016/j.ins.2014.02.091

7. Melin, P., Castillo, O.: Spatial and temporal spread of the COVID-19 pandemic using self organizing neural networks and a fuzzy fractal approach. Sustainability 13(15), 8295 (2021). https://doi.org/10.3390/su13158295

8. Barbounis, T.G., Theocharis, J.B.: Locally recurrent neural networks for wind speed prediction using spatial correlation. Inf. Sci. 177(24), 5775–5797 (2007). https://doi.org/10.1016/j.ins.2007.05.024

9. Melin, P., Sánchez, D., Castillo, O.: Genetic optimization of modular neural networks with fuzzy response integration for human recognition. Inf. Sci. 197, 1–19 (2012). https://doi.org/10.1016/j.ins.2012.02.027

10. Chacon, H.D., Kesici, E., Najafirad, P.: Improving financial time series prediction accuracy using ensemble empirical mode decomposition and recurrent neural networks. IEEE Access 8, 117133–117145 (2020). https://doi.org/10.1109/ACCESS.2020.2996981

11. Sarah, S., Mustakim, S., Novita, R., Rozanda, N.E.: Implementation of fuzzy c-means and self-organizing map for data clustering of palm oil. In: Proceedings of the 2023 International Seminar on Intelligent Technology and its Applications: Leveraging Intelligent Systems to Achieve Sustainable Development Goals, ISITIA 2023—Proceeding, pp. 444–449. Institute of Electrical and Electronics Engineers Inc. (2023). https://doi.org/10.1109/ISITIA59021.2023.10221173

12. Prakaisak, I., Wongchaisuwat, P.: Article hydrological time series clustering: a case study of telemetry stations in Thailand. Water **14**(13), 1–15 (2022). https://doi.org/10.3390/w14132095
13. Yao, J., Lu, B., Zhang, J.: Multi-step-ahead tool state monitoring using clustering feature-based recurrent fuzzy neural networks. IEEE Access **9**, 113443–113453 (2021). https://doi.org/10.1109/ACCESS.2021.3104668
14. Melin, P., Castillo, O.: An intelligent hybrid approach for industrial quality control combining neural networks, fuzzy logic and fractal theory. Inf. Sci. **177**(7), 1543–1557 (2007). https://doi.org/10.1016/j.ins.2006.07.022
15. Sánchez, D., Melin, P.: Modular neural networks for time series prediction using type-1 fuzzy logic integration. Stud. Comput. Intell. **601**, 141–154 (2015). https://doi.org/10.1007/978-3-319-17747-2_11
16. Wu, J.L., Lu, M., Wang, C.Y.: Forecasting metro rail transit passenger flow with multiple-attention deep neural networks and surrounding vehicle detection devices. Appl. Intell. **53**(15), 18531–18546 (2023). https://doi.org/10.1007/s10489-023-04483-x
17. Soto, J., Melin, P., Castillo, O.: Time series prediction using ensembles of ANFIS models with genetic optimization of interval type-2 and type-1 fuzzy integrators. Int. J. Hybrid Intell. Syst. **11**(3), 211–226 (2016). https://doi.org/10.3233/his-140196
18. Sotirov, S., Sotirova, E., Melin, P., Castilo, O., Atanassov, K.: Modular neural network preprocessing procedure with intuitionistic fuzzy InterCriteria analysis method. Adv. Intell. Syst. Comput. **12**, 175–186 (2016). https://doi.org/10.1007/978-3-319-26154-6_14
19. Pulido, M., Melin, P.: Comparison of genetic algorithm and particle swarm optimization of ensemble neural networks for complex time series prediction. In: Melin, P., Castillo, O., Kacprzyk, J. (eds) Recent Advances of Hybrid Intelligent Systems Based on Soft Computing. Studies in Computational Intelligence, Vol. 915, pp. 51–77. Springer, Cham (2021). https://doi.org/10.1007/978-3-030-58728-4_3
20. Ramirez, E., Melin, P., Prado-Arechiga, G.: Hybrid model based on neural networks, type-1 and type-2 fuzzy systems for 2-lead cardiac arrhythmia classification. Exp. Syst. Appl. **126**, 295–307 (2019). https://doi.org/10.1016/j.eswa.2019.02.035
21. Ontiveros-Robles, E., Castillo, O., Melin, P.: Towards asymmetric uncertainty modeling in designing general type-2 fuzzy classifiers for medical diagnosis. Exp. Syst. Appl. **183**, 115370 (2021). https://doi.org/10.1016/j.eswa.2021.115370
22. Rostam Niakan Kalhori, M., FazelZarandi, M.H.: A new interval type-2 fuzzy reasoning method for classification systems based on normal forms of a possibility-based fuzzy measure. Inf. Sci. **581**, 567–586 (2021). https://doi.org/10.1016/j.ins.2021.09.060.
23. Baskov, O.V., Noghin, V.D.: Type-2 fuzzy sets and their application in decision-making: general concepts. Sci. Tech. Inform. Process. **49**(5), 283–291 (2022). https://doi.org/10.3103/S014768822205001X
24. Xu, T.T., Qin, J.D.: A new representation method for type-2 fuzzy sets and its application to multiple criteria decision making. Int. J. Fuzzy Syst. **25**(3), 1171–1190 (2023). https://doi.org/10.1007/s40815-022-01432-7
25. Cheng, C.H., Chen, M.Y., Chang, J.R.: Linguistic multi-criteria decision-making aggregation model based on situational ME-LOWA and ME-LOWGA operators. Granul. Comput. **8**(1), 97–110 (2023). https://doi.org/10.1007/s41066-022-00316-3
26. Putra Perdana, M.Y., Fiade, A., Malik Matin, I.M.: Fuzzy multi-criteria decision making for optimization of housing construction financing. In: Proceedings of the 2021 6th International Conference on Informatics and Computing, ICIC 2021, pp. 1–5. Institute of Electrical and Electronics Engineers Inc. (2021). https://doi.org/10.1109/ICIC54025.2021.9632934
27. Chiao, K.P.: MCDM prioritization based on interval type 2 intuitionistic fuzzy sets ranking with parametric general graded mean integration representation. In: Proceedings of the 2022 International Conference on Fuzzy Theory and Its Applications, iFUZZY 2022, pp. 1–6. Institute of Electrical and Electronics Engineers Inc. (2022). https://doi.org/10.1109/iFUZZY55320.2022.9985226
28. Azzabi, L., Azzabi, D., Kobi, A.: The multi-criteria approach decision. In: International Series in Operations Research and Management Science, Vol. 300, pp. 1–23. Springer, New York (2020). https://doi.org/10.1007/978-3-030-57262-4_1

29. Zhao, S., Dong, Y., Martine, L., Pedrycz, W.: Analysis of ranking consistency in linguistic multiple attribute decision making: the roles of granularity and decision rules. IEEE Trans. Fuzzy Syst. **30**(7), 2266–2278 (2022). https://doi.org/10.1109/TFUZZ.2021.3078817

30. Xu, Y., Liu, S., Wang, J., Shang, X.: A novel two-stage TOPSIS approach based on interval-valued probabilistic linguistic q-rung orthopair fuzzy sets with its application to MAGDM problems. Eng. Appl. Artif. Intell. **116**, 105413 (2022). https://doi.org/10.1016/j.engappai.2022.105413

31. Gupta, A., Sharma, K.: Ranking of countries using world development indicators: a computational approach. In: Proceedings of the 2020 11th International Conference on Computing, Communication and Networking Technologies (ICCCNT), Kharagpur, India, pp. 1–4 (2020). https://doi.org/10.1109/ICCCNT49239.2020.9225403

32. Mathew, M., Chakrabortty, R.K., Ryan, M.J.: Selection of an optimal maintenance strategy under uncertain conditions: an interval type-2 fuzzy AHP-TOPSIS method. IEEE Trans. Eng. Manag. **69**(4), 1121–1134 (2022). https://doi.org/10.1109/TEM.2020.2977141

33. Akram, M., Khan, A., Luqman, A., Senapati, T., Pamucar, D.: An extended MARCOS method for MCGDM under 2-tuple linguistic q-rung picture fuzzy environment. Eng. Appl. Artif. Intell. **120**, 105892 (2023). https://doi.org/10.1016/j.engappai.2023.105892

34. Howlett, C.A., et al.: Same room: different windows? A systematic review and meta-analysis of the relationship between self-report and neuropsychological tests of cognitive flexibility in healthy adults. Clin. Psychol. Rev. **88**, 102061 (2021). https://doi.org/10.1016/j.cpr.2021.102061

35. Dheer, R.J.S., Lenartowicz, T.: Cognitive flexibility: Impact on entrepreneurial intentions. J. Vocat. Behav. **115**, 103339 (2019). https://doi.org/10.1016/j.jvb.2019.103339

36. Orakcı, Ş: Exploring the relationships between cognitive flexibility, learner autonomy, and reflective thinking. Think Skills Creat. **41**, 100838 (2021). https://doi.org/10.1016/j.tsc.2021.100838

Chapter 3
Problem Description of Prediction with Neural Networks

For several decades, decision-making support tools have offered assistance for knowing and providing access to information. The use and dissemination of the results of searches and analysis of information acquires a decisive strategic importance in organizational development.

The need to evaluate quality in all areas is a constant that depends on the approach taken to each situation. Broadly speaking, it could be measured through a group of indicators where it is established whether each of them is active or inactive, whether it is present or not, whether it is on or off.

Then we identify the indicators and their status, through an evaluation process we give a weight or priority to each indicator. The individual weights or values of each indicator can be combined to give a numerical or non-numerical value to the final evaluation.

Among the evaluable aspects we can mention economic indicators, sustainable development, and quality of life. In recent years, attention has been paid to the factors that directly impact well-being and how to evaluate it.

For this purpose, methodologies have been implemented to measure whether the basic needs of security, pleasant surroundings, access to digital and communications services are met. These indicators are being promoted worldwide, however there is a large gap between most countries.

It is very important that the used data is derived from public sources and supported by prestigious organizations, such as the World Bank Group, World Health Organization, among others. The selection of evaluation methods derives from the variables to be measured and the results to be obtained.

Once the indicators are obtained and evaluated, when combined, an index or criterion is formed, which is a guide to measure a country's capacity to achieve, take advantage of, respond to, or adapt to change.

© The Author(s), under exclusive license to Springer Nature Switzerland AG 2024   17
P. Melin et al., *Clustering, Classification, and Time Series Prediction by Using Artificial Neural Networks*, SpringerBriefs in Computational Intelligence,
https://doi.org/10.1007/978-3-031-71101-5_3

Among the criteria to be evaluated we can consider aspects of long-term future planning and governance strategies, talent management, industry strength, combating corruption and political structure to take advantage of the most recent opportunities presented by globalization.

The persistence and constant analysis of indicators and criteria worldwide seek to make improvements in institutional aspects, which, due to their origin, tend to change or modify slowly but their impact is broad coverage.

As mentioned in the previous chapters, decision makers are interested in looking for solutions that consider multiple criteria that arise from different sources of information and, in turn, prioritize these criteria based on their cost–benefit and quality measurement, in addition to considering the limitations or restrictions of complex problems.

Unlike decision-making in everyday life where it is possible to weight them from greater to lesser importance based on day-to-day experience, in organizational decision-making the predisposition is that each decision must be logical and rational and for the sake of meeting objectives and goals in a concise manner.

In both scenarios, decision makers must be able to adapt effectively through a set of cognitive and behavioral strategies in response to changes in the task or modification of environmental demands.

Although, they must deal with changes in the variables or attributes of the information, thus it is also necessary to consider the factor of ambiguity or implicit uncertainty present at the time of the decision, because this factor directly affects the desired results.

Therefore, it is essential to have a new bioinspired computational model that simulate the behavior of the cognitive functions of the brain when a person analyzes the information in their environment prior to deciding, which integrates multiple results from clustering, classification, and time series prediction.

This means that by using different intelligent hybrid methods we try to simulate the way in which a person groups and classifies data and infers what a future value could be for a given variable.

Thus, it can be used as a support tool for decision making with uncertainty both in organizational and government environments (Fig. 3.1).

Throughout this work, we assume that due to multiple internal or external factors, the timely integration of data is increasingly less frequent, which is why eight case studies are addressed where the analysis of historical information is considered as a support tool for decision-making.

3.1 Case Studies

Below we present the case studies developed in this research work, considering that as mentioned above, for decades the amount of data in organizations has increased rapidly and the time before making decisions has decreased considerably.

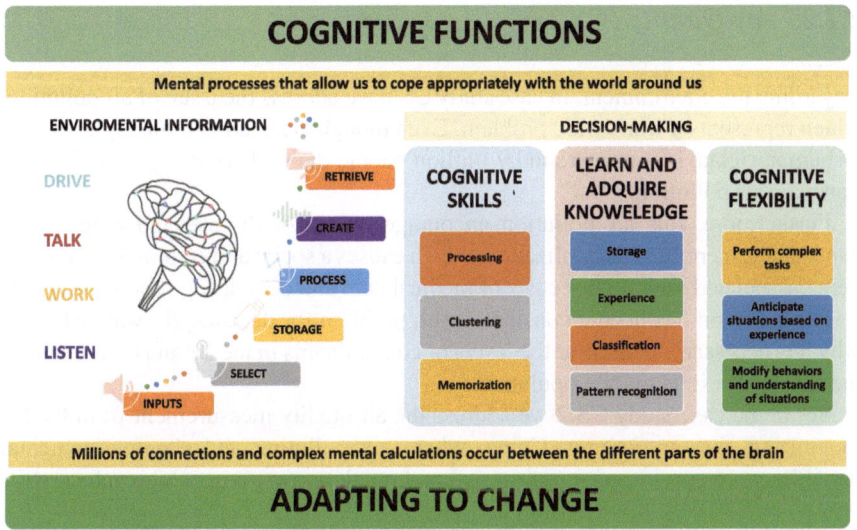

Fig. 3.1 Cognitive functions for decision-making

Highlighting uncertainty, changes in the environment and the growing need to consider multiple criteria prior to decision-making from a global or particular approach for each case study as appropriate.

3.1.1 Traffic Accidents (CS.1)

During recent decades, the problem of traffic accidents and injuries has been identified as a priority public health and development problem, which tests the way of managing the health sector in the public administration and undermining their ability to devote limited resources to other areas of need.

The purpose of the accident study is to plan and organize transportation and roads, for which annual information is produced on the accident rate of land transportation at the national level, at the level of federal entity and municipality.

This issue is addressed in this case study CS.1 corresponding to Land Traffic Accidents in Urban and Suburban Areas in Mexico data set (ATUS statistics), where traffic accident statistics are presented by municipality, state and nationally [1].

It consists of daily accident records for the 32 states of Mexico. We selected six attributes: state, year, month, day, gender, and age range of the driver. With these variables, clustering, classification, and prediction results were obtained.

3.1.2 Air Quality (CS.2)

Regarding the environment, in case study CS.2 we address the issue of air pollution, which represents a worldwide problem. Even though clean air is a fundamental part of human rights, however around 7 million people die each year due to ambient or domestic air pollution.

Furthermore, this air pollution encourages cases of disease and reduces life expectancy, increases hospitalization, which causes a social and economic impact by being associated with higher costs of medical care and decreasing productivity [2].

Air pollution has been a constant health problem for decades globally, which is why it is necessary to monitor the levels of contaminants in the air and communicate the obtained results to the population.

So, in the case study CS.2 we address the air quality measurement from twelve monitoring sites in Beijing, China, where six polluting and six environmental variables were considered for clustering and prediction experimental results.

3.1.3 Multiple Country Indicators (CS.3)

Global indicators are dynamic and sometimes the correlation is uncertain because they mainly depend on a combination of economic, social, and environmental factors. Therefore, in this case CS.3, we address these general aspects [3].

Because above, multiple time series as population, urban population, fine particles (PM2.5), carbon dioxide (CO_2), Covid19 cases, Covid19 deaths, for thirteen countries were considered for clustering and prediction criteria.

3.1.4 Multiple OECD Indicators (CS.4)

We deal with a constant challenge that arises in organizational environments, which is the combination of strategic indicators based on the results of time series analysis. One of the aspects is to delimit the membership of indicators in a particular environment, process, geographic region, or category [4, 5].

So, in the case study CS.4, we address the institutional indicators for member countries of the Organization for Economic Cooperation and Development (OECD): Population, OECD risk, Gross National Income (GNI), Inflation, Carbon dioxide (CO_2) emissions.

Finally, the relevance of using fuzzy integration to combine primary and secondary indicators and generate general indicators to create a proposal for decision-making is addressed for clustering and prediction criteria.

3.1.5 Global OECD Criteria (CS.5)

When it comes to knowing the level of development of a country, multiple measurements are frequently made to have a general overview. Among the aspects to consider are the measurements of indicators that impact the economy, the environment and society.

Generally, to measure the existing conditions in a country to ensure the well-being of its population, the total income of its residents is taken as a basis, which is represented by the GNI.

In this case study CS.5, we deal the evolution of some social and economic factors has contributed to the changes experienced by global components. Additionally, the constant changes in significant represent new challenges and opportunities for governments.

Thus, a modular perspective in a neural network model is addressed for population, and GNI time series prediction.

3.1.6 Consumption Levels LDCs (CS.6)

Likewise, in the case study CS.6 the clustering and prediction of the levels of consumption of Carbon dioxide emissions from liquid fuel and Renewable energy in Least Developed Countries (LDCs) is considered, also addresses matters relating to economic risk, which is a probability that measures the possible alterations.

As well as the generated uncertainty, and sometimes it could cause the impossibility of guaranteeing the compliance level with organizational goals and objectives, which is why for their administration they are frequently divided into multiple categories according to their consequences and impact.

3.1.7 Indicators of Latin American Countries (CS.7)

As part of continuous improvement processes, it is increasingly common for organizations to make historical records of information on economic and non-economic criteria, seeking to identify the behavior of a variable in different environments.

The need to separate components and groups them to reduce the number of criteria required for decision-making, complying with the achievement of goals and objectives.

Therefore, in the case study CS.7, the clustering and prediction of indicators of Latin American countries: Inflation, GNI, Population, Life expectancy, Labor force, Unemployment, GNI increase, Population increase, GNI per capita [6] are addressed.

3.1.8   World Development Indicators (CS.8)

Based on the behavior of multiple economic and non-economic variables during a given period, we will be able to identify key aspects in which improvement is viable.

In the case study CS.8, the clustering and prediction of World Development Indicators (WDI): Access to electricity, Birth rate, Death rate, Life expectancy at birth (female), Life expectancy at birth, Life expectancy birth (male), Population growth, Population, Population (female), %Population (female), Population (male), %Population (male), were considered.

We focus on treating data from a group of 208 countries to identify the significant similarities and discrepancies between them, as a strategy to compare the advances, setbacks or stagnation in the economic events that occurred on a given date.

3.2   Time Series Datasets

In Table 3.1 we show the list of the datasets (time series) used for each case study, where the code, name, periodicity, period, and code assigned to each time series (variable) are indicated. The frequency or periodicity can be Hourly (H_P), Daily (D_P) or Annually (A_P).

In Table 3.2 we can find the list of the main consulted sources during the preparation and development of the case studies addressed in this work, from where we obtained the datasets used in the multiple case studies presented previously. The data were classified by officially recognized international sources based on certain metrics.

In order to indicate the general aspects of this work, Table 3.3 lists the case studies addressed by this work, indicating their code, summary, number of instances or elements and the code of the selected time series.

For all the above, this chapter presents the aspects to be considered in our proposal regarding the problem to be solved, starting from the need to separate components and group them to reduce the number of criteria required for decision making.

As well as to predict future values and meet goals and objectives, this proposal contemplates a hybrid-hierarchical computational model that, broadly speaking, is a fuzzy integrator of multiple results of clustering, classification, and time series prediction.

It serves as a support tool for decision making with uncertainty in both an organizational and government environment using supervised and unsupervised neural networks and Type-1, Interval Type-2, and Type-2 fuzzy systems.

Table 3.1 Time series dataset

No	Time series code	Time series name	Periodicity	Period	Case study
1	TSC.01 [7]	ATUS accidents records	D_P	2012–2019	CS.1
2	TSC.02 [8]	Beijing multi-site air-quality data	H_P	2013–2017	CS.2
3	TSC.03 [9]	Number of COVID19 new confirmed cases and new deaths	D_P	2020–2022	CS.3
4	TSC.04 [10]	Urban population	A_P	1960–2020	CS.3
5	TSC.05 [11]	PM2.5 air pollution, mean annual exposure	A_P	2010–2017	CS.3
6	TSC.06 [12]	Carbon dioxide (CO_2) emissions	A_P	1960–2018 1990–2018 1994–2016	CS.3 CS.4 CS.6
7	TSC.07 [13]	Population	A_P	1960–2022 1960–2022 1990–2018 2006–2020 1960–2022 1960–2022	CS.3 CS.4 CS.4 CS.5 CS.7 CS.8
8	TSC.08 [14]	Inflation	A_P	1993–2020	CS.4
9	TSC.09 [15]	OECD country risk	A_P	1987–2020	CS.4
10	TSC.10 [16]	Gross national income (GNI)	A_P	2006–2020 2006–2020 1995–2021	CS.4 CS.5 CS.7
11	TSC.11 [17]	Renewable energy consumption	A_P	1990–2020	CS.6
12	TSC.12 [21]	Inflation	A_P	1990–2021	CS.7
13	TSC.13 [17]	Unemployment	A_P	1991–2021	CS.7
14	TSC.14 [17]	GNI per capita	A_P	1995–2021	CS.7
15	TSC.15 [17]	GDP per capita growth	A_P	1971–2021	CS.7
16	TSC.16 [17]	Life expectancy at birth	A_P	1960–2020	CS.7
17	TSC.17 [17]	Labor force	A_P	1990–2021	CS.7
18	TSC.18 [17, 18]	Population growth	A_P	1961–2021 1960–2022	CS.7 CS.8
19	TSC.19 [18]	Access to electricity	A_P	2000–2021	CS.8
20	TSC.20 [18]	Population (female)	A_P	1960–2022	CS.8
21	TSC.21 [18]	%Population (female)	A_P	1960–2022	CS.8
22	TSC.22 [18]	Population (male)	A_P	1960–2022	CS.8
23	TSC.23 [18]	%Population (male)	A_P	1960–2022	CS.8
24	TSC.24 [18]	Birth rate	A_P	1990–2021	CS.8
25	TSC.25 [18]	Death rate	A_P	1990–2021	CS.8

(continued)

Table 3.1 (continued)

No	Time series code	Time series name	Periodicity	Period	Case study
26	TSC.26 [18]	Life expectancy at birth (female)	A_P	2000–2021	CS.8
27	TSC.27 [18]	Life expectancy at birth	A_P	1990–2021	CS.8
28	TSC.28 [18]	Life expectancy birth (male)	A_P	1960–2021	CS.8

Table 3.2 List of dataset repositories

No	Logo	Dataset repository	Description	Established
1	INEGI	National Institute of Statistics and Geography (INEGI)	The INEGI is an autonomous public body responsible for disseminating information about the national territory of Mexico and its resources, population, and economy. It records and disseminates the characteristics to support decision making	1983
2	UCI	UCI Machine Learning Repository (UCI ML)	The UCI machine learning repository is a collection of databases frequently used for the empirical analysis of machine learning algorithms which has prestige among the research community	1987
3	OECD	Organisation for Economic Co-operation and Development (OECD)	The OECD is an international organization that is responsible for promoting the definition of better policies that help achieve a better life where opportunities are presented under equal conditions, in search of prosperity and well-being for all	1960
4		World Bank Group (WBG)	The World Bank Group is made up of five institutions that promote sustainable development, in addition to joining efforts to reduce poverty and increase shared prosperity, becoming one of the main sources of consultation and financing in the world for developing countries	1944
5		World Health Organization (WHO)	The WHO is responsible for guiding and promoting global guidelines, seeking to ensure that everyone everywhere can access and have the opportunity to live healthily	1948
6		United Nations (UN)	The UN is constantly transforming to evolve and be relevant to the global changes that occur every day	1949

Table 3.3 List of case studies

No	Case study code	Case study summary	Instances	Time series code
1	CS.1	**Traffic accidents** The Land Traffic Accidents in Urban and Suburban Areas in Mexico dataset (ATUS statistics) consists of daily accident records for the 32 states of Mexico and was created by the National Institute of Statistics and Geography (INEGI). We selected six attributes out of a total of 45: state, year, month, day, gender, and age range of the driver. With these variables, clustering, classification, and prediction results were obtained The purpose of the accident study is to plan and organize transportation and roads, for which annual information is produced on the accident rate of land transportation at the national level, at the level of federal entity and municipality	32	TS01
2	CS.2	**Air quality** We selected the Beijing dataset from the UCI machine learning repository, which consists of the hourly air pollutant record from a total of 12 nationally controlled air quality monitoring sites of the municipal environmental monitoring center of Beijing. To monitor the air, information records are used for a given period and the current measurement and the future measurement of a polluting index or variable are notified, which may be the cause of a health or respiratory system problem, such as particles suspension (PM2.5). We calculated the monthly average of the variables PM10, PM25, NO_2, O_3, SO_2 and CO for clustering and for prediction the monthly and daily average of the PM25 variable was calculated. Also, we considered these environmental variables TEMP, PRES, DEWP, RAIN, WD, WSPM and monitoring site name for clustering	12	TS02
3	CS.3	**Multiple country indicators** We aim at identifying associations between different time series: population, urban population, particulate matter (PM2.5), carbon dioxide (CO_2) and COVID 19 using an intelligent hybrid computational model. it should be noted that no preprocessing was carried out prior to its use. Each time series have six attributes each (country name, country code, indicator name, indicator code, value indicator, year)	13	TS03, TS04, TS05, TS06, TS07

(continued)

Table 3.3 (continued)

No	Case study code	Case study summary	Instances	Time series code
4	CS.4	**Multiple OECD indicators** The relevance of using fuzzy integration to combine primary and secondary indicators and generate general indicators to create a proposal for decision-making is addressed. Therefore, it is proposed to use a Type-2 fuzzy integration for a set of multivariate time series: country risk classification, inflation, population, and OECD gross national income (GNI). It should be noted that time series have six attributes each (country name, country code, indicator name, indicator code, value indicator, year) and no data preprocessing was performed. Also, High Income OECD members countries have been not classified; it means belongs to category 0	38	TS06, TS07, TS08, TS09, TS10
5	CS.5	**Global OECD criteria** When it comes to knowing the level of development of a country, multiple measurements are frequently made to have a general overview. Among the aspects to consider are the measurements of indicators that impact the economy, the environment and society Generally, to measure the existing conditions in a country to ensure the well-being of its population, the total income of its residents is taken as a basis, which is represented by the GNI We propose a method to predict population and gross national income time series. It should be noted that both series have six attributes each (country name, country code, indicator name, indicator code, value indicator, year) and no data preprocessing was performed	38	TS07, TS10
6	CS.6	**Consumption levels LDCs** A significant factor in the negative impact of the operation of a developing country is the change in global economic conditions, combined with the possibility of the materialization of one or many of the risks that threaten its economic environment. Through the analysis of time series, it is possible to identify patterns or similarity in the information over time, as well as to predict future values, for which it is necessary to have technological tools that contribute to the timely integration of the results of the study carried out. Thus, our proposal consists of a method for clustering and prediction levels of consumption of Carbon dioxide emissions from liquid fuel and Renewable energy in developing countries time series For all records of Least Developed Countries (LDCs) dataset, we have six attributes for each of the countries (type indicator, code indicator, name, code, value, year). Also, no data preprocessing was performed	46	TS06, TS11

(continued)

Table 3.3 (continued)

No	Case study code	Case study summary	Instances	Time series code
7	CS.7	**Indicators of Latin American countries** Like most people, decision makers gain their skills when faced with challenges to structure complex problems and gain experience. Which makes the use of tools to integrate indicators through a hierarchical arrangement of multiple criteria increasingly common. The need to separate components and groups them to reduce the number of criteria required for decision-making, complying with the achievement of goals and objectives. So, we proposed a model for classification and prediction of country indicators such as inflation, unemployment, population, population growth, labor force, among others. For the dataset of Inflation, GNI, GNI per capita, and GDP per capita we only used the last value of the time series for Cuba (2020–2021) and Venezuela (2015–2021), because data for these periods were unavailable. Also, no data preprocessing was performed	20	TS07, TS10, TS12, TS13, TS14, TS15, TS16, TS17, TS18
8	CS.8	**World development indicators** Based on the behavior of multiple economic and non-economic variables during a given period, we will be able to identify key aspects in which improvement is viable. In this case we are referring to a sample of 208 countries, which present similarities and contrasts, with which it will be possible to visualize economic progress, stagnation and setbacks on a given date. The variables (time series): Access to electricity, Birth rate, Death rate, Life expectancy at birth (female), Life expectancy at birth, Life expectancy birth (male), Population growth, Population, Population (female), %Population (female), Population (male), %Population (male), were considered for clustering and prediction criteria	208	TS07, TS18, TS19, TS20. TS21, TS22, TS23, TS24, TS25, TS26, TS27, TS28

References

1. Ramirez, M., Melin, P.: Clustering and prediction of time series for traffic accidents using a nested layered artificial neural network model. In: Castillo, O., Melin, P. (eds.) New Perspectives on Hybrid Intelligent System Design based on Fuzzy Logic, Neural Networks and Metaheuristics. Studies in Computational Intelligence, Vol. 1050, pp. 37–46. Springer, Cham (2022). https://doi.org/10.1007/978-3-031-08266-5_3
2. Ramírez, M., Melin, P.: Multiple neural networks for clustering and prediction of the particulate matter (PM2.5): a case study of Beijing. In: Kahraman, C., Sari, I.U., Oztaysi, B., Cebi, S., Cevik Onar, S., Tolga, A.Ç. (eds.) Intelligent and Fuzzy Systems. INFUS 2023. Lecture Notes in Networks and Systems, Vol. 759, pp. 507–514. Springer, Cham (2023). https://doi.org/10.1007/978-3-031-39777-6_60

3. Ramírez, M., Melin, P.: A new interval type-2 fuzzy aggregation approach for combining multiple neural networks in clustering and prediction of time series. Int. J. Fuzzy Syst. **25**(3), 1077–1104 (2023). https://doi.org/10.1007/s40815-022-01426-5

4. Ramirez, M., Melin, P.: A decision-making approach based on multiple neural networks for clustering and prediction of time series. In: Castillo, O., Melin, P. (eds.) Hybrid Intelligent Systems Based on Extensions of Fuzzy Logic, Neural Networks and Metaheuristics. Studies in Computational Intelligence, Vol. 1096, pp. 3–14 (2023). https://doi.org/10.1007/978-3-031-28999-6_1

5. Ramirez, M., Melin, P.: A new perspective for multivariate time series decision making through a nested computational approach using type-2 fuzzy integration. Axioms **12**(4), 385 (2023). https://doi.org/10.3390/axioms12040385

6. Ramírez, M., Melin, P., Castillo, O.: Interval type-3 fuzzy aggregation for hybrid-hierarchical neural classification and prediction models in decision-making. Axioms **12**(10), 906 (2023). https://doi.org/10.3390/axioms12100906

7. Ground traffic accidents in urban and suburban areas. http://en.www.inegi.org.mx/programas/accidentes/#Documentation. Accessed 22 Sept 2020

8. Zhang, S., Guo, B., Dong, A., He, J., Xu, Z., Chen, S.X.: Cautionary tales on air-quality improvement in Beijing. Proceedi. Royal Soc. A Math. Phys. Eng. Sci. **473**(2205), 1–14 (2017). https://doi.org/10.1098/rspa.2017.0457

9. Ritchie, H., et al.: Coronavirus pandemic (COVID-19). Published online at OurWorldIn-Data.org. https://ourworldindata.org/coronavirus. Accessed 08 Feb 2022

10. The World Bank Data: Urban population. https://data.worldbank.org/indicator/SP.URB.TOTL. Accessed 08 Feb 2022

11. The World Bank Data: PM2.5 air pollution, mean annual exposure (micrograms per cubic meter). https://data.worldbank.org/indicator/EN.ATM.PM25.MC.M3. Accessed 08 Feb 2022

12. The World Bank Data: CO_2 emissions (kt). https://data.worldbank.org/indicator/EN.ATM.CO2 E.KT. Accessed 08 Feb 2022

13. The World Bank Data: Population total. https://data.worldbank.org/indicator/SP.POP.TOTL. Accessed 08 Feb 2022

14. The World Bank Data: Inflation. https://data.worldbank.org/indicator/FP.CPI.TOTL.ZG. Accessed 08 Feb 2022

15. OECD: Country Risk Classifications of the Participants to the Arrangement on Officially Supported Export Credit. https://www.oecd.org/trade/topics/export-credits/documents/cre-crc-current-english.pdf. Accessed 05 April 2022

16. The World Bank Data: GNI per capita. https://data.worldbank.org/indicator/NY.GNP.MKT P.CD. Accessed 22 March 2022

17. The World Bank Data: Renewable energy consumption total. https://data.worldbank.org/indicator/EG.FEC.RNEW.ZS. Accessed 10 July 2023

18. The World Bank Data: World Development Indicators. https://datatopics.worldbank.org/world-development-indicators/. Accessed 6 May 2023

Chapter 4
Methodology for Prediction with Neural Networks

In general, our proposal consists of combining several methods to create a computational model that simulates the cognitive functioning of the human brain; focusing on the brain processes implicit in the mental routine that the human being performs to decision-making, where each mental process is directed towards the solution of a specific task (Fig. 4.1).

The methods considered include neural networks, fuzzy systems, time series and the theory of cognitive flexibility focused on the decision-making process. Through which it is intended to partially simulate how cognitive skills work.

This means that they work simultaneously, are recursive, maintain a hierarchy, are activated, or deactivated based on experience, evolve based on the complexity of the problem to be solved and experience acquired.

Each mental process is aimed at solving a specific task, for which mental routines are created to solve tasks in a sequential and orderly or sporadic manner.

Cognitive functions make it possible to react to the constant incentives of the environment, causing a process of adaptation to change to occur when new situations arise and even with the experience acquired, it is not possible to solve problems in the known way.

The central idea of this work to partially simulate cognitive functions is that from an input dataset prepared for this model, four levels are available that operate sequentially or independently, this means that the data enters directly at each level, and it is possible to obtain results of certain specific methods of the model (Fig. 4.2).

The four levels proposed in the model are described below:

- **First level:** it is composed of a set of supervised and unsupervised neural networks, which are used to perform clustering, classification, and prediction tasks. These neural networks work individually (monolithically), and collaboratively (combined, ensemble, modular, among others).

P. Melin et al., *Clustering, Classification, and Time Series Prediction by Using Artificial Neural Networks*, SpringerBriefs in Computational Intelligence, https://doi.org/10.1007/978-3-031-71101-5_4

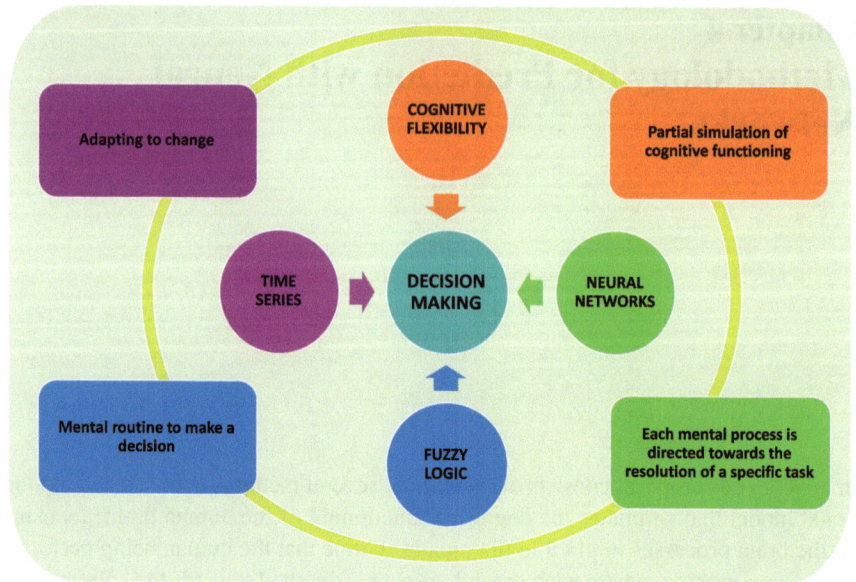

Fig. 4.1 General aspects of the proposed model

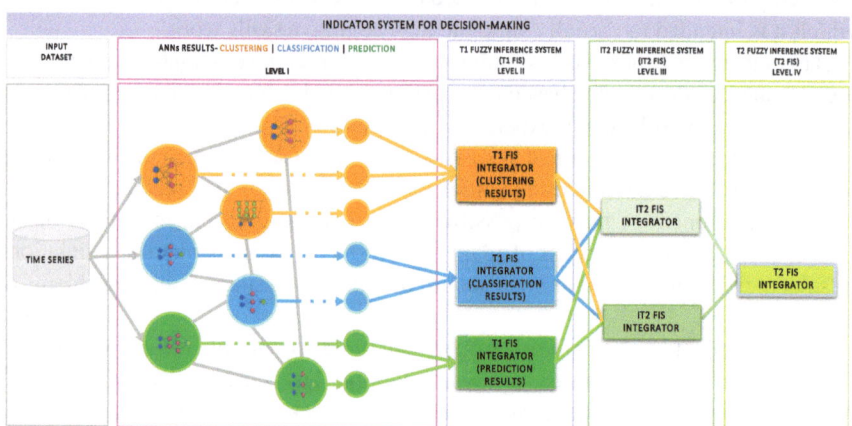

Fig. 4.2 Proposed method

Considering the problem to be solved, we can choose which type of neural network to select and whether they will be combined to work simultaneously or first the results of a group of networks will be obtained and based on these results other neural networks will be executed.

- **Second level:** it is composed of a set of Type-1 fuzzy systems, which are responsible for integrating the obtained results using neural networks (Level I) or the data entered directly by the user, based on their operational need.
- **Third level:** it is composed of a set of interval Type-2 fuzzy systems, which are responsible for integrating the obtained results using Type-1 fuzzy systems (Level II) or the data entered directly by the user, based on their operational need.
- **Fourth level:** it consists of a Generalized Type-2 fuzzy system, which acts as an integrator of the obtained results by Interval Type-2 fuzzy systems (Level III) or of the data entered directly by the user based on their operational need.

Since the proposed model is scalable, this means that, based on the required application by the user, it can be used completely or through independent levels. In addition, to the fact that each level can be composed of multiple methods configured particularly for the solution of the assigned task.

Besides, levels can operate interdependently, since results obtained at a certain level can be used at a previous or subsequent level.

We tested the model with multiple case studies, seeking to test different methods for each clustering, classification, and prediction task, and we also carried out tests seeking to obtain results from the combination of these tasks.

In addition to entering data directly at the exclusive levels of neural networks (Level I) or at the levels corresponding to fuzzy systems that act as indicator integrators (Levels II–IV). The visual description of the models used for each study is presented below.

During the development of the work carried out to perform clustering tasks we used Self-organizing map (SOM) and Competitive neural networks.

In the case of prediction tasks, we use Long short-term memory (LSTM), Nonlinear autoregressive neural network (NAR) and Nonlinear autoregressive network with exogenous inputs (NARX).

Finally, for classification tasks we use Feedforward neural network (FNN), Type-1, Interval Type-2, and Generalized Type-2 fuzzy inference systems.

Also, later the particularities of each fuzzy model are presented as appropriate for each case study developed in this research work.

Both Type-1 and Type-2 fuzzy inferences systems utilized to aggregate the artificial neural networks results, are formed by two to four inputs and one output in all cases, they are Mamdani type, from eight to twenty-seven fuzzy rules, and centroid defuzzification.

It should be noted that the membership function (MF) parameters were manually tested, until obtaining these parameter values. The MFs are trapezoidal, Gaussian, and triangular. The linguistic values of each variable are generally described in each fuzzy system model.

For Type-1 fuzzy systems (Fig. 4.3) the linguistic values of each variable are: Very few (VFWW), Few (FWW), Many (MNN), and Too many (TMNN).

For Interval Type-2 fuzzy systems (Fig. 4.4) the linguistic values of each variable are: Low (LWW), Medium (MDD), and High (HGG).

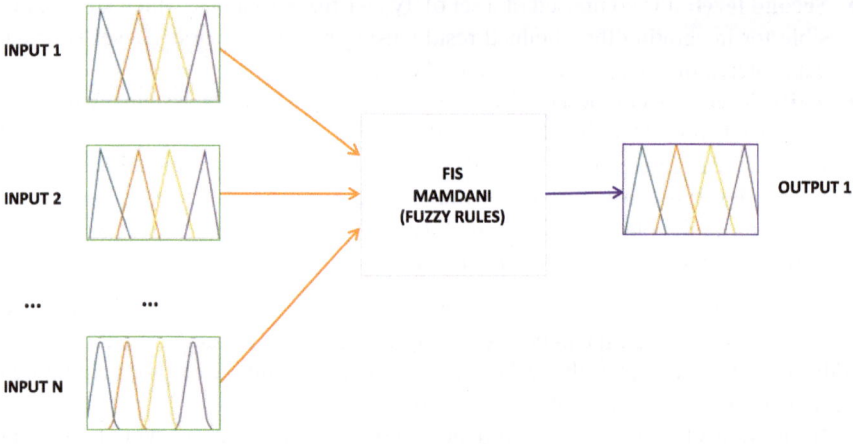

Fig. 4.3 Type-1 fuzzy system model

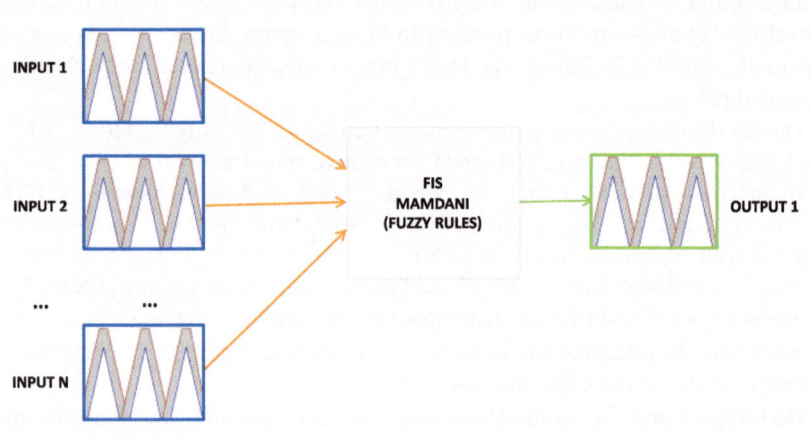

Fig. 4.4 Interval Type-2 fuzzy system model

In the case of Generalized Type-2 fuzzy systems (Fig. 4.5) the linguistic values of each variable are: Low (LWW), Medium (MDD), and High (HGG).

In general, the fuzzy rules used in Type-1 Mamdani FIS are shown below in the Table 4.1.

For both Interval Type-2 and Generalized Type-2 fuzzy system, the fuzzy rules used are shown in Table 4.2.

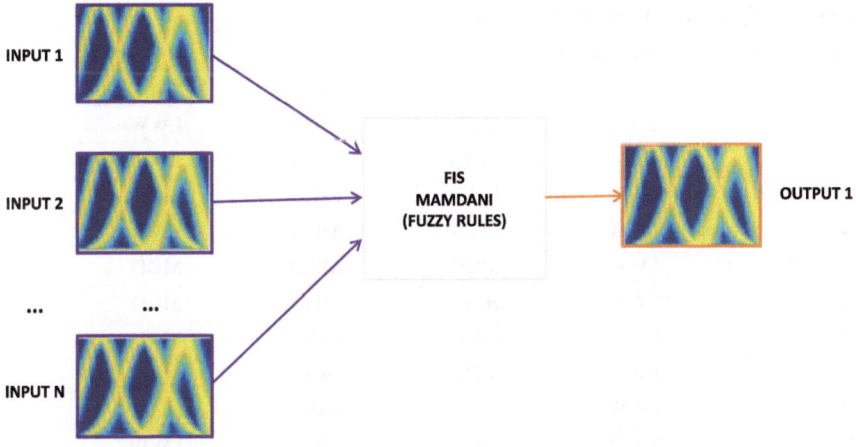

Fig. 4.5 Generalized Type-2 fuzzy inference system model

Table 4.1 Type-1 FIS Mamdani fuzzy rules

| Fuzzy rules | Antecedents | | | | Consequent |
	INPUT_1	INPUT_2	INPUT_3	INPUT_4	OUTPUT_1
1	INPUT 1 is VFWW	INPUT 2 is VFWW	INPUT 3 is VFWW	INPUT 4 is VFWW	OUTPUT 1 is VFWW
2	INPUT 1 is FWW	INPUT 2 is FWW	INPUT 3 is FWW	INPUT 4 is VFWW	OUTPUT 1 is FWW
3	INPUT 1 is VFWW	INPUT 2 is FWW	INPUT 3 is VFWW	INPUT 4 is FWW	OUTPUT 1 is FWW
4	INPUT 1 is FWW	INPUT 2 is VFWW	INPUT 3 is FWW	INPUT 4 is VFWW	OUTPUT 1 is FWW
5	INPUT 1 is FWW	INPUT 2 is MNN	INPUT 3 is FWW	INPUT 4 is MNN	OUTPUT 1 is MNN
6	INPUT 1 is MNN	INPUT 2 is FWW	INPUT 3 is MNN	INPUT 4 is FWW	OUTPUT 1 is MNN
7	INPUT 1 is MNN	INPUT 2 is TMNN	INPUT 3 is MNN	INPUT 4 is TMNN	OUTPUT 1 is TMNN
8	INPUT 1 is TMNN	INPUT 2 is MNN	INPUT 3 is TMNN	INPUT 4 is MNN	OUTPUT 1 is TMNN
9	INPUT 1 is MNN	INPUT 2 is MNN	INPUT 3 is MNN	INPUT 4 is TMNN	OUTPUT 1 is TMNN
10	INPUT 1 is TMNN	INPUT 2 is TMNN	INPUT 3 is TMNN	INPUT 4 is TMNN	OUTPUT 1 is TMNN

Table 4.2 Type-2 FIS Mamdani fuzzy rules

Fuzzy rules	Antecedents			Consequent
	Input_1	Input_2	Input_3	Output_1
1	LWW	LWW	LWW	LWW
2	LWW	MDD	LWW	LWW
3	LWW	HGG	LWW	MDD
4	LWW	LWW	MDD	LWW
5	LWW	MDD	MDD	MDD
6	LWW	HGG	MDD	MDD
7	LWW	LWW	HGG	MDD
8	LWW	MDD	HGG	MDD
9	LWW	HGG	HGG	HGG
10	MDD	LWW	LWW	LWW
11	MDD	MDD	LWW	MDD
12	MDD	HGG	LWW	MDD
13	MDD	LWW	MDD	MDD
14	MDD	MDD	MDD	MDD
15	MDD	HGG	MDD	HGG
16	MDD	LWW	HGG	MDD
17	MDD	MDD	HGG	HGG
18	MDD	HGG	HGG	HGG
19	HGG	LWW	LWW	MDD
20	HGG	MDD	LWW	MDD
21	HGG	HGG	LWW	HGG
22	HGG	LWW	MDD	MDD
23	HGG	MDD	MDD	HGG
24	HGG	HGG	MDD	HGG
25	HGG	LWW	HGG	HGG
26	HGG	MDD	HGG	HGG
27	HGG	HGG	HGG	HGG

4.1   Case Study CS.1

In case study CS.1, the input dataset consists of accidents records time series, which enter to Level I to perform classification and clustering tasks. Regarding first level, we use a FNN for classification into three classes, and competitive neural networks to clustering into four classes.

Also, regarding the prediction tasks, we use LSTM neural network which is an upgraded version of Recurrent Neural Network (RNN). For second level, we use two Type-1 fuzzy systems as integrators of competitive neural networks results (Fig. 4.6).

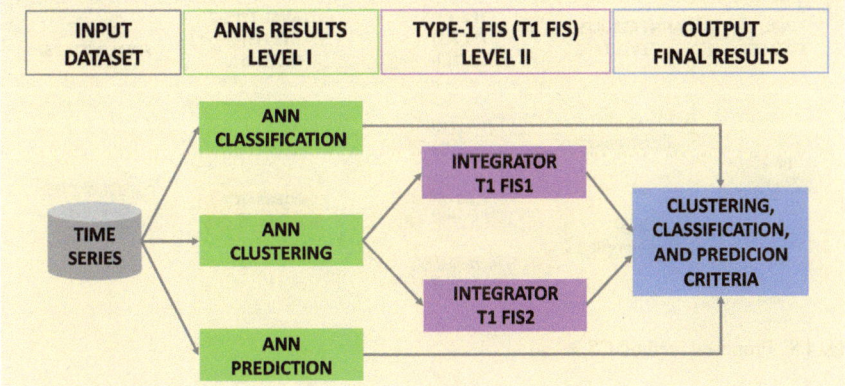

Fig. 4.6 Proposed method CS.1

4.2 Case Study CS.2

For case study CS.2, we use in the first level a SOM and competitive neural networks to clustering into four classes and, we use NAR for time series prediction. Also, for second level we use multiple Type-1 fuzzy systems as integrators of clustering and prediction neural networks results (Fig. 4.7).

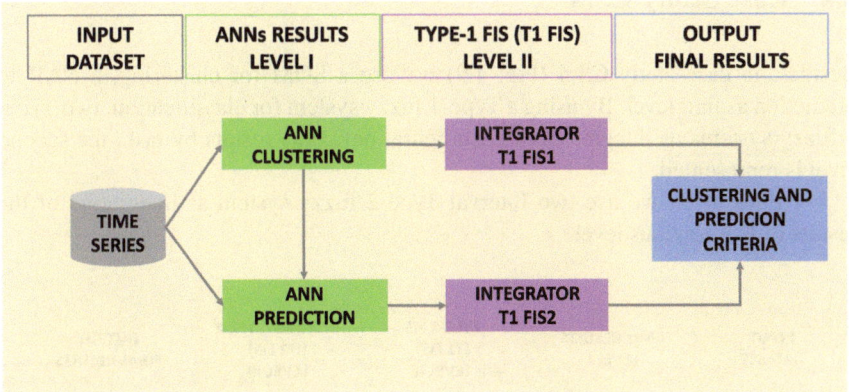

Fig. 4.7 Proposed method CS.2

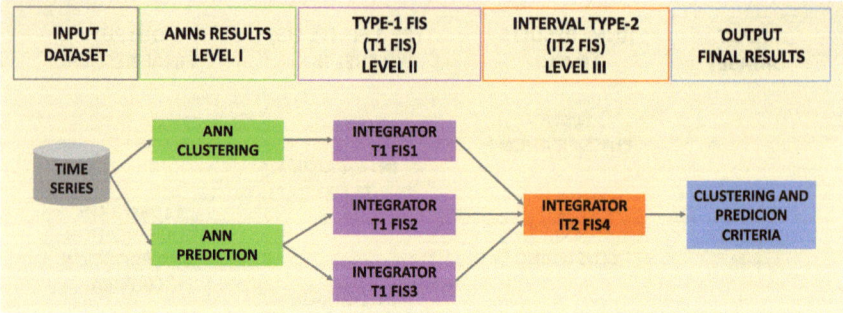

Fig. 4.8 Proposed method CS.3

4.3   Case Study CS.3

Likewise, in case study CS.3 (Fig. 4.8), we use a SOM for clustering, a NAR and a NARX to prediction as Level I. By using three Type-1 fuzzy systems as integrators of each neural networks results by task the Level II is represented. For Level III we use and Interval Type-2 fuzzy system as an integrator of the results of the previous level.

4.4   Case Study CS.4

Besides, in case study CS.4 (Fig. 4.9) we use a SOM for clustering, a NAR to prediction as first level. By using a Type-1 fuzzy system for classification, two Type-1 fuzzy systems as integrators of each neural networks results by task, the second level is represented.

For third level, we use two Interval Type-2 fuzzy system as integrators of the results of the previous level.

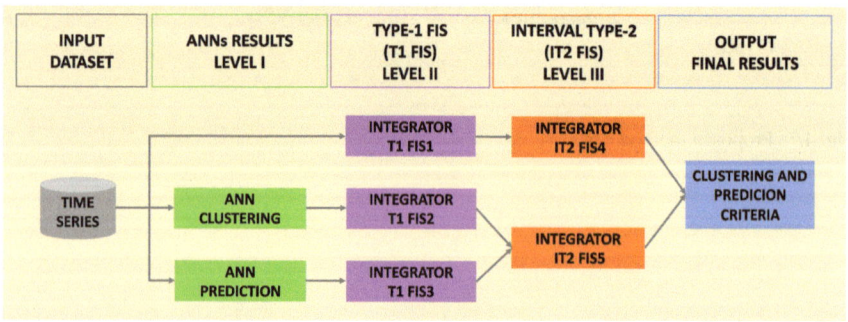

Fig. 4.9 Proposed method CS.4

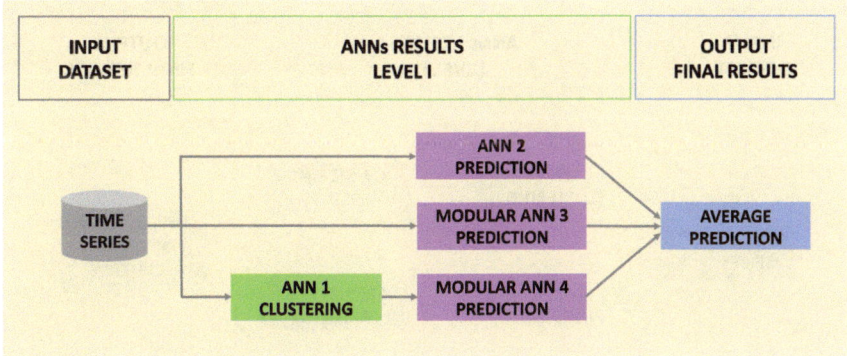

Fig. 4.10 Proposed method CS.5

4.5 Case Study CS.5

In the CS5 case study (Fig. 4.10), where we first use a SOM for clustering into four classes, and then we use a set of NAR to predict the complete time series and each of the cluster.

Finally, we perform an integration of the neural networks prediction results by average. It should be noted that we use a modular perspective in a neural network model (monolithic or modular).

4.6 Case Study CS.6

For the case study CS6, we use a monolithic NAR to predict, and we use a SOM for clustering into four classes as first level (Fig. 4.11). Finally, clustering and prediction criteria results are obtained.

4.7 Case Study CS.7

Furthermore, in case study CS.7, we use a NAR for prediction, a set of SOM for clustering into four classes, as Level I. By using four Type-1 fuzzy systems as integrators of each neural networks results by task the Level II is represented.

Also, for Level III we use and Interval Type-2 fuzzy system as an integrator of the results of the previous level. In the last level, we perform an integration using a Generalized Type-2 fuzzy system. Finally, clustering and prediction criteria results are obtained (Fig. 4.12).

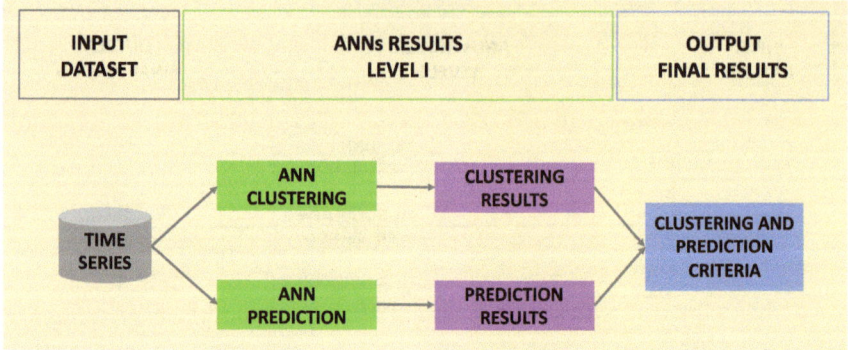

Fig. 4.11 Proposed method CS.6

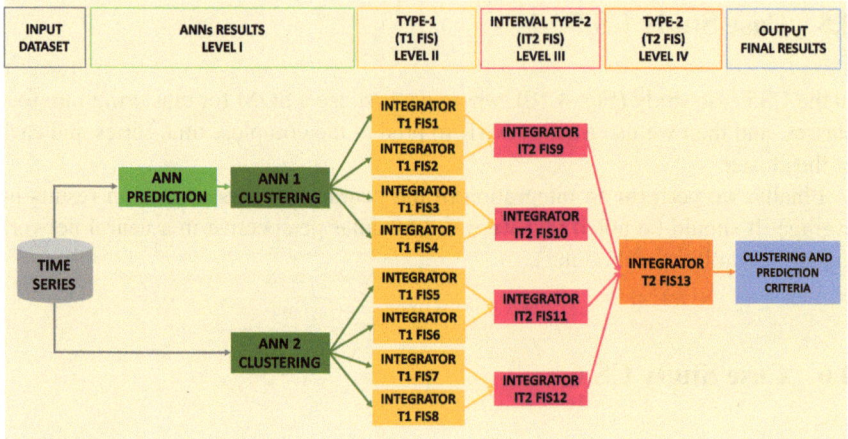

Fig. 4.12 Proposed method CS.7

4.8 Case Study CS.8

Similarly, in case study CS.8 (Fig. 4.13), for Level I we use a NAR for prediction, a SOM and competitive neural networks for clustering into six classes. By using six Type-1 fuzzy systems as integrators of each neural networks results by task the Level II is represented.

In Level III we use seven Interval Type-2 fuzzy systems as an integrator of the results of the previous level. Finally, for level 4 with these results of Type-1 and Interval Type-2, we use a Generalized Type-2 fuzzy inference system which operate as integrator of the results. So, we expect to achieve the best global result for this problem.

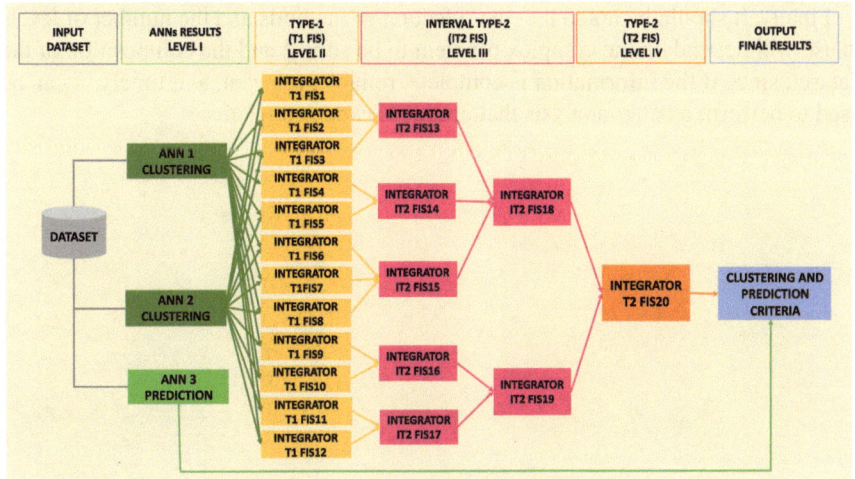

Fig. 4.13 Proposed method CS.8

The list of methods used for each of the case studies mentioned above are presented below for each of the tasks (Table 4.3).

Based on the above, in this chapter we can highlight that this proposal arises from the need to create a computational model that simulates the cognitive behavior of the brain during the decision-making process, considering performing clustering, classification, and prediction of multivariate time series individually or simultaneous.

For which we separate the tasks into four phases or levels, with the objective of presenting a sequence and highlighting the use of supervised and unsupervised neural networks at Level I, and Type-1, Interval Type-2 and Generalized Type-2 fuzzy systems at Levels II–IV.

Table 4.3 Description of methods used by tasks/CS

No	Tasks	Methods	CS.1	CS.2	CS.3	CS.4	CS.5	CS.6	CS.7	CS.8
1	Clustering	SOM	✔	✔	✔			✔	✔	✔
		Competitive neural network	✔	✔		✔	✔			✔
2	Prediction	LSTM	✔							
		NAR	✔	✔	✔	✔	✔	✔	✔	✔
		NARX			✔					
3	Classification	FNN	✔							
		T1 FIS	✔	✔	✔	✔			✔	✔
		IT2 FIS			✔	✔			✔	✔
		T2 FIS							✔	✔

Finally, it should be noted that the selection of methods and the number of levels to use will depend on the complex problem to be solved and the composition of the dataset, since if the information is complete, reliable, relevant, and timely, it can be used to perform a better analysis that produces expected results.

Chapter 5
Results of Prediction with Neural Networks

For the case studies where the neural networks experiments were use, we performed 30 executions for each experiment.

In the case of supervised neural networks, because good results were obtained in previous experimentation, we divide the original sequence of the time series in 70% of dataset for training, 15% validation and 15% for testing. It should be noted that the sequential order of each time series remained.

Also, the relative percentage of Root Mean Square Error (%RMSE) was used to measure the prediction performance of each neural network and Mean Squared Error (MSE) was used to measure the classification performance of FNN.

Below we present the obtained results for each case study.

5.1 Case Study CS.1

By using an FNN with 10 neurons in the hidden layer, we classify accidents per day at the national level into three classes: fatal, non-fatal and damage only. The best result obtained corresponds to 99.95% classification and the worst result corresponds to 92.35%. Finally, a classification average of 98.50% was obtained.

Based on the variable monthly number of accidents (VC1) recorded in the period 2013–2018, the states that recorded similar traffic accident statistics were grouped into four classes: Very few (VFF), Few (FWW), Many (MNN) or Too many accidents (TMNN) by using competitive neural networks.

As well, based on the variables corresponding to the driver: age range (VC2 and VC5), age range women (VC3 and VC6), age range men (VC4 and VC7) similar clusters were formed respectively.

It is about analyzing whether we group the states by taking the information from the period 2013–2018 for VC1, VC2, VC3 and VC4 variables (Table 5.1), and then observing how many clusters remain similar when we group them based on VC5, VC6 and VC7 variables corresponding to 2019 information only (Table 5.2).

Once we identify the states that belong to each class based on the results of the unsupervised neural networks, we prepare the inputs of the Type-1 fuzzy systems for fuzzy integration whose purpose is to identify the membership of each state to a certain group.

We noticed that when the results of the variables VC1, VC5, VC6 and VC7 were integrated, a greater number of elements were separated compared to the variables VC1, VC2, VC3 and VC4 (Table 5.3).

Besides, we made the prediction of the number of accidents by state (VC1), age range (VC2), age range women (VC3), age range men (VC4) of the driver, based on the dataset of the 32 states respectively (Table 5.4).

Table 5.1 CS.1 clusters by state 2013–2018

Variable	VFF	FWW	MNN	TMNN
VC1	28	2	1	1
VC2	19	11	1	1
VC3	29	1	1	1
VC4	18	12	1	1

Table 5.2 CS.1 clusters by state 2019

Variable	VFF	FWW	MNN	TMNN
VC5	18	12	1	1
VC6	17	10	4	1
VC7	23	7	1	1

Table 5.3 CS.1 type-1 fuzzy system results

FIS	Type-1	Type-1
Inputs	VC1 VC2 VC3 VC4	VC1 VC5 VC6 VC7
Output	FIS1 T1	FIS2 T1
VFF	28	17
FWW	1	11
MNN	2	3
TMNN	1	1

Table 5.4 CS.1 LSTM prediction of the number accidents variables

Variable	Average %RMSE	Best %RMSE	Worst %RMSE
VC1	0.001099408	0.000892990	0.001426313
VC2	0.002031722	0.001478331	0.002679007
VC3	0.002571199	0.001437401	0.003503766
VC4	0.002084991	0.001511632	0.002845864

Table 5.5 CS.1 LSTM prediction of number accidents by clusters VC1

Cluster	Elements	Average %RMSE	Best %RMSE	Worst %RMSE
VFF	28	0.001089330	0.000966223	0.001405463
FWW	2	0.004298581	0.001711493	0.010981270
MNN	1	0.015911502	0.013330322	0.019791860
TMNN	1	0.004718022	0.002490341	0.007786377

Then, we separate the data corresponding to each state using the classes formed by the competitive neural networks. Then, with these data segments we train the LSTM neural network to predict the variable VC1 (Table 5.5).

5.2   Case Study CS.2

As part of the obtained results, it was possible to form four classes (CC1, CC2, CC3, CC4) by using competitive neural networks, consisting of the monitoring sites (MNST) that recorded monthly concentrations of six contaminant variables. For which we take as a basis the average monthly concentration of variable PM10, PM25, NO_2, O_3, CO, SO_2 for a period between 2013 and 2017.

In addition, we carry out the prediction of the PM25 variable for each of the clusters formed. We calculate the Average, Best and Worst percentage of the relative RMSE for the obtained results as shown below (Table 5.6).

Table 5.6 CS.2 prediction for clusters of average daily PM2.5 (multiple variables)

Class	MNST	Size	Average %RMSE	Best %RMSE	Worst %RMSE
CC1	7	10,227	0.00003446	0.00003407	0.00003527
CC2	2	2922	0.00020571	0.00020172	0.00021161
CC3	2	2922	0.00021293	0.00020702	0.00022527
CC4	1	1461	0.00059876	0.00047015	0.00369635

Based on the average monthly concentration of the variable PM25 by using competitive neural networks we classify the monitoring sites into four classes. Subsequently, we made the prediction of the PM25 variable for each of the clusters formed (Table 5.7).

Afterwards, for the clusters formed based on the information of the six polluting variables, the prediction of the monthly average of the PM25 variable was made. (Table 5.8).

Also, based on the clusters formed using competitive neural networks we made the prediction of the average monthly variable PM25 variable for each cluster (Table 5.9).

Similarly, we made the prediction of the average daily (Table 5.10) and average monthly (Table 5.11) PM25 variables.

Additionally, we integrate a sample of 32 elements of the outputs of the unsupervised neural networks to obtain the final cluster: Low (LWW), Medium (MDD), High (HGG) for each element (Table 5.12).

Table 5.7 CS.2 prediction for clusters of average daily PM2.5 (individual variables)

Class	MNST	Size	Average %RMSE	Best %RMSE	Worst %RMSE
CC1	2	2922	0.00019933	0.00018991	0.00024125
CC2	3	4383	0.00012389	0.00012092	0.00013315
CC3	4	5844	0.00010090	0.00009914	0.00010343
CC4	3	4383	0.00010871	0.00010702	0.00011159

Table 5.8 CS.2 prediction for clusters of monthly PM2.5 (multiple variables)

Class	MNST	Size	Average %RMSE	Best %RMSE	Worst %RMSE
CC1	7	336	0.00050199	0.00046543	0.00074532
CC2	2	96	0.00342633	0.00230991	0.00632787
CC3	2	96	0.00377913	0.00307685	0.00633747
CC4	1	48	0.00678047	0.00508235	0.01450819

Table 5.9 CS.2 prediction for clusters of monthly PM2.5 (individual variables)

Class	MNST	Size	Average %RMSE	Best %RMSE	Worst %RMSE
CC1	2	96	0.00273402	0.00234010	0.00472917
CC2	3	144	0.00202317	0.00157779	0.00410329
CC3	4	192	0.00110263	0.00095638	0.00161259
CC4	3	144	0.00183348	0.00146168	0.00321809

Table 5.10 CS.2 prediction results for six average daily pollution variables

Variable	Average %RMSE	Best %RMSE	Worst %RMSE
PM10	0.00001861	0.00001846	0.00001888
PM25	0.00002079	0.00002051	0.00002092
NO_2	0.00002248	0.00002238	0.00002277
O_3	0.00001709	0.00001703	0.00001717
CO	0.00002412	0.00002374	0.00002434
SO_2	0.00003233	0.00003203	0.00003371

Table 5.11 CS.2 prediction for six average monthly pollution variables

Variable	Average %RMSE	Best %RMSE	Worst %RMSE
PM10	0.000305745	0.000298708	0.000329358
PM25	0.000307320	0.000293227	0.000341818
NO_2	0.000523055	0.000501873	0.000581513
O_3	0.000328529	0.000316985	0.000395431
CO	0.000412227	0.000398324	0.000439063
SO_2	0.000626912	0.000533747	0.002499245

Table 5.12 CS.2 type-1 fuzzy system results

FIS	Type-1	Type-1
Inputs	CC_PM25 CC_PM10 CC_SO_2 CC_NO_2	SUM_PM25_PM10 SUM_SO_2_NO_2
Output	FIS1 T1	FIS2 T1
LWW	28	16
MDD	2	16
HGG	0	0

5.3 Case Study CS.3

Firstly, by using SOM networks, we classify the variables Population (V1), Urban population (V2), PM25 (V3), CO_2 (V4), Covid19 deaths (V5), Covid19 cases (V6) into four classes based on the average total annual records (Table 5.13).

For the prediction of four variables, we used a NARX neural network with 10 neurons in the hidden layer (Table 5.14).

Also, we used a NAR neural network with 10 neurons in the hidden layer to make the prediction of the four variables V1, V2, V4 and V6 (Table 5.15).

Subsequently, we classified using a Mamdani Type-1 fuzzy inference system with four inputs and one output, based on the variables V1, V2, V4 and V6, the level reached for each of the countries according to the clusters made by the neural

Table 5.13 CS.3 clusters formed based on individual variables (multiple countries)

Variable	Cluster 1 (C1)	Cluster 2 (C2)	Cluster 3 (C3)	Cluster 4 (C4)
V1	8	2	2	1
V2	7	3	1	2
V3	4	4	3	2
V4	9	2	1	1
V5	9	2	1	1
V6	10	1	1	1

Table 5.14 CS.3 prediction of multiple pollution variables using NARX

Primary variable	Secondary variable	Average %RMSE	Best %RMSE	Worst %RMSE
V1	V2	0.000096798	0.000004321	0.001512516
V2	V1	0.000137810	0.000007525	0.002548414
V4	V1	0.031689279	0.005474208	0.078434301
V6	V5	0.000076041	0.000018428	0.000166901

Table 5.15 CS.3 prediction of individual pollution variables using NAR

Variable	Average %RMSE	Best %RMSE	Worst %RMSE
V1	0.000390816	0.000039335	0.002322634
V2	0.001407223	0.000071357	0.009816944
V4	0.001486522	0.000649318	0.003356711
V6	0.010909516	0.005269325	0.036478969

networks SOM (Very few (VFF), few (FWW), many (MNN), too many (TMNN) records (Table 5.16).

Next, the level of increase of a variable was calculated by subtracting from the final value (prediction made by the neural network) the initial value (original data) of

Table 5.16 CS.3 results of type-1 FIS (country level)

FIS	Type-1
Inputs	V1
	V2
	V4
	V6
Output	FIS1 T1
VFF	0
FWW	11
MNN	1
TMNN	1

Table 5.17 CS.3 results of type-1 and type-2 FIS (country level)

FIS	Type-1	Type-1	Interval type-2
Inputs	Increase V1 Increase V2	Increase V4 Increase V6	FIS1 T1 FIS2 T1 FIS3 T1
Output	FIS2 T1	FIS3 T1	FIS4 T1
LW	6	3	5
MM	7	9	8
HG	0	1	0

the time series. The obtained result was then divided by the starting value (original data).

Once the level of increase was obtained for each of the four variables: V1, V2, V4 and V6, we used a second Mamdani Type-1 fuzzy inference system with two inputs and one output, to classify the level of increase of V1 and V2 variables.

Subsequently, a third Mamdani Type-1 fuzzy inference system with two inputs and one output was used to classify the level of increase of the variables V4 and V6.

Finally, we use a Type-2 fuzzy inference system to classify the level of general increase of the variables V1, V2, V4 and V6, based on the outputs of the Type-1 fuzzy inference systems (Table 5.17).

5.4 Case Study CS.4

We carried out some clustering experiments where the population and dioxide carbon emissions time series were grouped into four classes (CC1, CC2, CC3, CC4) by using competitive neural networks (Table 5.18).

Next, we present the prediction results by using the complete dataset of the population and carbon dioxide emissions time series (Table 5.19).

Table 5.18 CS.4 list of total countries per clusters (CC1, CC2, CC3, CC4)

Cluster Id	Population	CO_2
CC1	25	5
CC2	10	12
CC3	2	9
CC4	1	12

Table 5.19 CS.4 time series prediction (population, carbon dioxide emissions)

Variable	Average %RMSE	Best %RMSE	Worst %RMSE
Population	0.00699930	0.00488868	0.00884851
CO_2	0.00044016	0.00031031	0.00079133

Table 5.20 CS.4 time series prediction by clusters (population)

Variable	Average %RMSE	Best %RMSE	Worst %RMSE
CC1	0.00131000	0.00036386	0.00250869
CC2	0.00097472	0.00046697	0.00225286
CC3	0.00042760	0.00005231	0.00379156
CC4	0.00083594	0.00003595	0.00333030

Table 5.21 CS.4 time series prediction by clusters (carbon dioxide emissions)

Variable	Average %RMSE	Best %RMSE	Worst %RMSE
CC1	0.00087009	0.00033204	0.00179916
CC2	0.00053669	0.00031860	0.00089270
CC3	0.00035440	0.00014393	0.00066555
CC4	0.00133659	0.00074783	0.00203253

Then, we use the data corresponding to the results of the competitive neural networks to identify the data from each cluster to train a set of NAR neural networks for prediction tasks. The prediction results for each cluster of the population time series (Table 5.20) and carbon dioxide emissions time series (Table 5.21) are shown below.

Afterwards, by using a Mamdani Fuzzy Type-1 inference system with four inputs and one output, we classify each country based on the GNI variable. The indicator level reached for each of the countries according to the corresponding classification for every calendar year.

Seeking to integrate the results respect to the class assigned to each country: Lower income (L), Lower Middle income (LM), Upper Middle income (UM), and High income (H) into a new classification label (Table 5.22).

We first subtract the immediate previous value of the time series from the initial value. After obtained the result it is divided by the initial value.

Subsequently, we classify the variables of the level of increase in the population (V1) and the GNI (V2). Then, a third Mamdani Type-1 fuzzy inference system was used to classify the level of increase of inflation (V3) and OECD country risk (V4).

Finally, once the classification results of the level of increase of the variables is obtained by using three Type-1 fuzzy systems, these results are integrated by using a

Table 5.22 CS.4 output first type-1 FIS classification GNI

FIS	Type-1
Inputs	GNI
Output	FIS1 T1
L	0
LM	0
UM	10
H	28

Table 5.23 CS.4 results of type-1 and type-2 classification (country level)

FIS	Type-1	Type-1	Interval Type-2
Inputs	Increase V1 Increase V2	Increase V3 Increase V4	FIS1 T1 FIS2 T1 FIS3 T1
Output	FIS2 T1	FIS3 T1	FIS4 IT2
LW	16	38	2
MM	22	0	36
HG	0	0	0

Table 5.24 CS.4 total countries per clusters of time series

Cluster Id	GNI	Population
CC1	32	26
CC2	5	10
CC3	1	1
CC4	0	1

Type-2 fuzzy system, where the three inputs correspond to the classification of each of the Type-1 fuzzy systems.

For the purpose of obtaining an output that represents the final classification of the level of increase of the variables: Low (LW), Medium (MM) or High (HH), as appropriate (Table 5.23).

5.5   Case Study CS.5

Firstly, we use competitive neural networks to form four classes (CC1, CC2, CC3, CC4). So, the obtained results for the GNI, population, time series are shown in Table 5.24.

For prediction we use a modular neural network, where the first module consists of a NAR (NAR1), the second module of a modular NAR (Modular NAR1) and the third module of a modular NAR (Modular NAR2 Cluster).

In this case as input for each module receives a segment of the data (formed by a competitive network). Also, the final integration of the obtained results by averaging the results of each module can be seen in Tables 5.25 and 5.26.

5.6   Case Study CS.6

For prediction tasks of the CO_2 emissions from liquid fuel consumption and renewable energy consumption time series several NAR are used.

Table 5.25 CS.5 prediction of GNI time series

Method	Average %RMSE	Best %RMSE	Worst %RMSE
NAR1	0.001400470	0.000707375	0.002611694
Modular NAR1	0.001421085	0.000718226	0.002698286
Modular NAR2 Cluster	0.009458815	0.001858983	0.022013062
Final prediction (integration)	0.004093457	0.001094861	0.009107681

Table 5.26 CS.5 prediction of population time series

Method	Average %RMSE	Best %RMSE	Worst %RMSE
NAR1	0.000180752	0.00001776	0.000423866
Modular NAR1	0.000168653	0.00005415	0.000385271
Modular NAR2 Cluster	0.000834394	0.00003572	0.004684953
Final prediction (integration)	0.000394600	0.00003588	0.001831363

Highlighting that the results for prediction are obtained using the complete time series and with the advantage that there is also prediction by class or group, which allows additional analyzes to be carried out, as appropriate.

Also, the use of SOM neural networks is to find clusters based on the CO_2 emissions from liquid fuel consumption and renewable energy consumption recorded in a period for each of the time series.

The results of the clustering experiments for the CO_2 emissions from liquid fuel consumption and renewable energy consumption time series are shown below, where clusters are identified with the labels CC1, CC2, CC3 and CC4 (Table 5.27).

From this point we show the prediction results for the time series of the CO_2 emissions from liquid fuel consumption and the renewable energy consumption time series (Table 5.28).

Table 5.27 CS.6 total countries per clusters of time series

Cluster Id	CO_2	Renewable energy
CC1	5	8
CC2	8	25
CC3	7	6
CC4	26	7

Table 5.28 CS.6 prediction of consumption time series

Variable	Average %RMSE	Best %RMSE	Worst %RMSE
Emissions CO_2_LFC	0.00016189	0.00006062	0.00034015
Renewable_EC	0.00011226	0.00005931	0.00019339

Table 5.29 CS.6 prediction of CO_2 emissions from liquid fuel consumption by clusters

Variable	Average %RMSE	Best %RMSE	Worst %RMSE
Emissions CO_2_LFC C1	0.00233798	0.00096964	0.00714024
Emissions CO_2_LFC C2	0.00092878	0.00034326	0.00212455
Emissions CO_2_LFC C3	0.00086339	0.00042849	0.00261138
Emissions CO_2_LFC C4	0.00017354	0.00007444	0.00036776

Table 5.30 CS.6 prediction of renewable energy consumption time series by clusters

Variable	Average %RMSE	Best %RMSE	Worst %RMSE
Renewable_EC C1	0.00064560	0.00027734	0.00129060
Renewable_EC C2	0.000119083	0.00007074	0.00019758
Renewable_EC C3	0.002868994	0.00091248	0.005101734
Renewable_EC C4	0.000562614	0.000307864	0.001428718

Next, we can observe the prediction results for the clusters formed by the SOM neural networks, where we use the data corresponding to each group to train the NAR neural network. The above applies to both time series: CO_2 emissions from liquid fuel consumption series (Table 5.29) and renewable energy consumption time series (Table 5.30).

5.7 Case Study CS.7

It should be mentioned that to make the prediction of the historical data we use NAR for the variables (time series): Inflation (V1), Unemployment (V2), Population (V3), Population growth (V4), GNI (V5), GNI per capita (V6), GDP per capita growth (V7), Life expectancy at birth (V8), Labor force (V9). The prediction results are shown in Table 5.31.

By using SOM the clustering of the nine variables was carried out first based on the historical data in four classes (CC1, CC2, CC3,CC4) as shown in Table 5.32 and second on the prediction of historical data in four classes (CC1, CC2, CC3, CC4) as shown in Table 5.33.

In the first instance, we obtained the classification of the prediction and the historical values of the nine variables by using unsupervised neural networks.

So, the obtained results were integrated by using a Type-1 fuzzy system, firstly the classification based on the prediction of the future values of each time series (Table 5.34) and later the classification made with the historical values was obtained (Table 5.35).

The integration of the Type-1 fuzzy systems results by using interval Type-2 fuzzy systems are presented (Table 5.36). For each of the countries the classification

Table 5.31 CS.7 time series prediction (NAR)

Variable	Average %RMSE	Best %RMSE	Worst %RMSE
V1	0.00079592	0.00047190	0.00228534
V2	0.00070178	0.00041320	0.00113990
V3	0.00012504	0.00003087	0.00043363
V4	0.00022320	0.00016006	0.00034506
V5	0.00320891	0.00124069	0.00803200
V6	0.00240148	0.00043670	0.04217167
V7	0.00394129	0.00249697	0.00613884
V8	0.00003375	0.00001639	0.00005301
V9	0.00046956	0.00008781	0.00125127

Table 5.32 CS.7 clustering results of time series (SOM)

Variable	Cluster 1 (CC1)	Cluster 2 (CC2)	Cluster3 (CC3)	Cluster 4 (CC4)
V1	1	16	1	2
V2	6	7	6	1
V3	4	14	1	1
V4	4	13	2	1
V5	4	14	1	1
V6	8	6	4	2
V7	17	1	1	1
V8	6	2	7	5
V9	4	14	1	1

Table 5.33 CS.7 list of clusters from prediction results (SOM)

Variable	Cluster 1 (CC1)	Cluster 2 (CC2)	Cluster3 (CC3)	Cluster 4 (CC4)
V1	2	16	1	1
V2	7	9	1	3
V3	2	16	1	1
V4	2	8	8	2
V5	3	15	1	1
V6	3	7	5	5
V7	10	5	3	2
V8	11	7	1	1
V9	4	14	1	1

Table 5.34 CS.7 classification results of the prediction values (type-1 FIS)

FIS	Type-1			
Inputs	V1 V4	V2 V9	V3 V8	V5 V6 V7
Outputs (country results)	FIS1 T1	FIS2 T1	FIS3 T1	FIS4 T1
VFF	0	0	0	0
FWW	6	7	7	5
MNN	14	13	13	15
TMNN	0	0	0	0

Table 5.35 CS.7 classification results of the time series (type-1 FIS)

FIS	Type-1			
Inputs	V1 V4	V2 V9	V3 V8	V5 V6 V7
Outputs (country results)	FIS5 T1	FIS6 T1	FIS7 T1	FIS8 T1
VFF	0	0	0	0
FWW	7	6	6	7
MNN	13	14	14	13
TMNN	0	0	0	0

obtained through the Type-2 FIS showed different results, because that depends on each of the four inputs (which are the outputs of the set of Type-1 fuzzy systems).

Finally, the integration of the results of the Interval Type-2 by using a Generalized Type-2 are presented (Table 5.37). For each of the countries the classification obtained through the Generalized Type-2 FIS showed similar results, because that depends on each of the four inputs (which are the outputs of the set of Interval Type-2 FIS).

Table 5.36 CS.7 type-2 FIS integration results for type-1 FIS outputs

FIS	Interval type-2			
Inputs	FIS1 T1 FIS3 T1	FIS2 T1 FIS4 T1	FIS6 T1 FIS8 T1	FIS7 T1 FIS5 T1
Outputs (country results)	FIS9 IT2	FIS10 IT2	FIS11 IT2	FIS12 IT2
LW	0	0	0	0
MM	7	5	1	9
HG	13	15	19	11

Table 5.37 CS.7T generalized type-2 fis for integration of IT2 FISs outputs

FIS	Generalized type-2
Inputs	FIS 9 IT2 FIS 10 IT2 FIS 11 IT2 FIS 12 IT2
Output (country results)	FIS13 T2
LW	0
MM	20
HG	0

5.8 Case Study CS.8

For this case study we use the classification of seven regions of the World Bank for the selected sample of countries (Table 5.38).

Firstly, we present the obtained results from the clustering using SOM for the variables: Access to electricity (TSS1), Birth rate (TSS2), Death rate (TSS3), Life expectancy at birth female (TSS4), Life expectancy at birth (TSS5), Life expectancy birth male (TSS6), Population growth (TSS7), Population (TSS8), Population female (TSS9), %Population female (TSS10), Population male (TSS11), and %Population male (TSS12) respectively.

For all regions obtained results by using competitive neural networks are shown in Table 5.39, where the total countries by each class (RC1C, RC2C, RC3C, RC4C, RC5C and RC6C) are described. Also, the results by SOM are shown in Table 5.40, where the total countries by each class (RC1S, RC2S, RC3S, RC4S, RC5S and RC6) are presented.

Also, for each of the regions, first the obtained results by using competitive ANNs (regions) are shown in Table 5.41 (R1), Table 5.42 (R2), Table 5.43 (R3), Table 5.44 (R4), Table 5.45 (R5), Table 5.46 (R6), and Table 5.47 (R7).

Then, those of the SOM (regions) are shown in Table 5.48 (R1), Table 5.49 (R2), Table 5.50 (R3), Table 5.51 (R4), Table 5.52 (R5), Table 5.53 (R6), and Table 5.54 (R7).

Table 5.38 CS.8 list of total countries by region

No	Region code	Region name	Total countries
1	R1	East Asia and Pacific	35
2	R2	Europe and Central Asia	53
3	R3	Latin America and Caribbean	41
4	R4	Middle East and North Africa	20
5	R5	North America	3
6	R6	South Asia	8
7	R7	Sub-Saharan Africa	48

Table 5.39 CS.8 clustering of WDI time series by using competitive ANNs

Variable	RC1C	RC2C	RC3C	RC4C	RC5C	RC6C
TSS1	15	111	18	33	14	17
TSS2	31	35	38	32	40	32
TSS3	25	36	38	40	38	31
TSS4	28	52	41	42	30	15
TSS5	29	39	41	25	40	34
TSS6	29	40	21	40	43	35
TSS7	36	36	37	33	36	30
TSS8	206	2	0	0	0	0
TSS9	196	10	2	0	0	0
TSS10	39	41	34	16	38	40
TSS11	206	2	0	0	0	0
TSS12	38	41	34	17	39	39

Table 5.40 CS.8 clustering of WDI time series by using SOM

Variable	RC1S	RC2S	RC3S	RC4S	RC5S	RC6S
TSS1	18	112	21	22	17	18
TSS2	26	26	42	21	33	60
TSS3	2	16	68	22	64	36
TSS4	32	63	40	30	42	1
TSS5	27	50	66	25	39	1
TSS6	32	44	50	44	37	1
TSS7	60	57	28	57	5	1
TSS8	35	140	11	8	2	12
TSS9	37	147	7	2	14	1
TSS10	98	10	75	4	19	2
TSS11	33	140	13	8	2	12
TSS12	101	10	72	4	19	2

With the results of the ANNs and SOM, we formed the two inputs of a Type-1 fuzzy system, where once these two results were integrated, the final class of each of the variables per country was obtained. So, we presented the accumulated elements per region (Table 5.55).

Now, with the purpose of integrating the 12 classes previously integrated with the Type-1 fuzzy systems into five classes, we prepared the inputs for five Type-1 fuzzy systems, where there were two to three inputs and one output Table 5.56.

Afterwards we are integrating the 12 classes previously integrated into 5 classes, but now we used two to three input and one output variables in an Interval Type-2 fuzzy system (Table 5.57).

Table 5.41 CS.8 clustering of WDI time series (R1) by using competitive ANNs

Variable	R1C1C	R1C2C	R1C3C	R1C4C	R1C5C	R1C6C
TSS1	4	16	2	9	3	1
TSS2	0	7	7	8	4	9
TSS3	1	6	11	13	4	0
TSS4	0	7	7	8	12	1
TSS5	6	6	5	1	8	9
TSS6	4	6	1	7	8	9
TSS7	3	2	8	3	18	1
TSS8	34	1	0	0	0	0
TSS9	32	2	1	0	0	0
TSS10	13	8	2	6	3	3
TSS11	34	1	0	0	0	0
TSS12	13	8	2	6	3	3

Table 5.42 CS.8 clustering of WDI time series (R2) by using competitive ANNs

Variable	R2C1C	R2C2C	R2C3C	R2C4C	R2C5C	R2C6C
TSS1	0	51	0	2	0	0
TSS2	0	13	3	4	32	1
TSS3	0	11	1	8	12	21
TSS4	0	18	9	24	2	0
TSS5	0	11	10	0	23	9
TSS6	0	11	0	9	23	10
TSS7	0	27	3	19	4	0
TSS8	53	0	0	0	0	0
TSS9	51	2	0	0	0	0
TSS10	5	2	18	1	20	7
TSS11	53	0	0	0	0	0
TSS12	5	2	18	1	20	7

Then, we integrate the five obtained results using two interval and two Generalized Type-2 FIS, with the purpose of obtaining two criteria (Table 5.58).

Below we present the comparison of the obtained results using Type-1, and Interval Type-2 FIS (Table 5.59).

In Table 5.60 we show the comparison of the obtained results with the use of Interval Type-2, and Generalized Type-2 FIS.

The results of the prediction of the future values of each variable were obtained using a NAR neural network are shown below (Table 5.61).

Table 5.43 CS.8 clustering of WDI time series (R3) by using competitive ANNs

Variable	R3C1C	R3C2C	R3C3C	R3C4C	R3C5C	R3C6C
TSS1	0	24	1	14	2	0
TSS2	0	11	14	12	1	3
TSS3	0	12	12	12	4	1
TSS4	1	13	18	6	2	1
TSS5	1	14	17	0	3	6
TSS6	1	16	0	15	3	6
TSS7	3	6	14	9	6	3
TSS8	41	0	0	0	0	0
TSS9	39	2	0	0	0	0
TSS10	6	13	7	0	5	10
TSS11	41	0	0	0	0	0
TSS12	6	13	7	0	5	10

Table 5.44 CS.8 clustering of WDI time series (R4) by using competitive ANNs

Variable	R4C1C	R4C2C	R4C3C	R4C4C	R4C5C	R4C6C
TSS1	1	15	0	3	1	0
TSS2	1	2	7	6	1	3
TSS3	0	1	13	5	1	0
TSS4	1	8	7	2	2	0
TSS5	1	7	5	0	3	4
TSS6	1	6	0	5	6	2
TSS7	1	0	5	1	1	12
TSS8	20	0	0	0	0	0
TSS9	20	0	0	0	0	0
TSS10	5	3	0	7	0	5
TSS11	20	0	0	0	0	0
TSS12	5	3	0	7	0	5

We separated the data corresponding to R1 classified using the Generalized Type-2 fuzzy system, and the results of the prediction of the future values of each variable were obtained using a NAR neural network (Table 5.62).

Therefore, this chapter presents the obtained results in the experimentation of these eight case studies, which highlights the use of supervised neural networks to perform classification and prediction tasks, and unsupervised neural networks to perform clustering tasks.

Also, Type-1, Interval Type-2, and Generalized Type-2 fuzzy systems to perform classification tasks and to integrate the obtained results.

Table 5.45 CS.8 clustering of WDI time series (R5) by using competitive ANNs

Variable	R5C1C	R5C2C	R5C3C	R5C4C	R5C5C	R5C6C
TSS1	0	3	0	0	0	0
TSS2	0	1	0	0	2	0
TSS3	0	3	0	0	0	0
TSS4	0	1	0	2	0	0
TSS5	0	0	0	0	3	0
TSS6	0	0	0	0	3	0
TSS7	0	1	0	1	1	0
TSS8	3	0	0	0	0	0
TSS9	2	1	0	0	0	0
TSS10	0	2	0	0	1	0
TSS11	3	0	0	0	0	0
TSS12	0	2	0	0	1	0

Table 5.46 CS.8 clustering of WDI time series (R6) by using competitive ANNs

Variable	R6C1C	R6C2C	R6C3C	R6C4C	R6C5C	R6C6C
TSS1	2	0	0	3	3	0
TSS2	1	0	5	1	0	1
TSS3	1	1	1	1	4	0
TSS4	1	2	0	0	5	0
TSS5	1	1	1	0	0	5
TSS6	1	1	0	1	0	5
TSS7	1	0	2	0	2	3
TSS8	7	1	0	0	0	0
TSS9	5	2	1	0	0	0
TSS10	4	2	0	2	0	0
TSS11	7	1	0	0	0	0
TSS12	3	2	0	3	0	0

Different sets of time series were also considered to analyze and evaluate the obtained results using the selected methods and observe how they behaved when different historical data were included.

For which throughout these case studies, a variety of environmental, economic, and demographic variables, among others, were considered, seeking to develop the proposed model in a comprehensive and complete way.

Finally, a study was also carried out on the importance of selecting levels to be used within the proposed model, according to the case study analyzed and what would be the best method to use according to the complex problem to be solved.

Table 5.47   CS.8 clustering of WDI time series (R7) by using competitive ANNs

Variable	R7C1C	R7C2C	R7C3C	R7C4C	R7C5C	R7C6C
TSS1	8	2	15	2	5	16
TSS2	29	1	2	1	0	15
TSS3	23	2	0	1	13	9
TSS4	25	3	0	0	7	13
TSS5	20	0	3	24	0	1
TSS6	22	0	20	3	0	3
TSS7	28	0	5	0	4	11
TSS8	48	0	0	0	0	0
TSS9	47	1	0	0	0	0
TSS10	6	11	7	0	9	15
TSS11	48	0	0	0	0	0
TSS12	6	11	7	0	10	14

Table 5.48   CS.8 clustering of WDI time series (R1) by using SOM

Variable	R1C1S	R1C2S	R1C3S	R1C4S	R1C5S	R1C6S
TSS1	3	16	4	7	4	1
TSS2	0	10	6	3	7	9
TSS3	0	0	18	1	6	10
TSS4	8	7	11	0	8	1
TSS5	5	8	10	0	11	1
TSS6	0	6	10	8	10	1
TSS7	5	2	9	19	0	0
TSS8	4	23	2	2	1	3
TSS9	5	24	2	1	3	0
TSS10	17	0	7	0	11	0
TSS11	4	23	2	2	1	3
TSS12	18	0	6	0	11	0

Table 5.49 CS.8 clustering of WDI time series (R2) by using SOM

Variable	R2C1S	R2C2S	R2C3S	R2C4S	R2C5S	R2C6S
TSS1	0	51	0	2	0	0
TSS2	0	1	7	0	3	42
TSS3	0	13	12	0	27	1
TSS4	0	19	10	0	24	0
TSS5	0	26	17	0	10	0
TSS6	0	11	13	24	5	0
TSS7	0	42	5	6	0	0
TSS8	4	40	3	1	0	5
TSS9	6	40	1	0	6	0
TSS10	13	9	29	0	2	0
TSS11	4	40	3	1	0	5
TSS12	13	9	29	0	2	0

Table 5.50 Clustering of WDI time series (R3) by using SOM

Variable	R3C1S	R3C2S	R3C3S	R3C4S	R3C5S	R3C6S
TSS1	0	25	1	9	6	0
TSS2	0	6	19	0	11	5
TSS3	0	0	24	0	6	11
TSS4	3	21	11	0	6	0
TSS5	1	7	25	0	8	0
TSS6	0	20	14	3	4	0
TSS7	5	12	7	15	2	0
TSS8	5	32	2	1	0	1
TSS9	5	34	1	0	1	0
TSS10	25	1	15	0	0	0
TSS11	5	32	2	1	0	1
TSS12	25	1	15	0	0	0

Table 5.51 CS.8 clustering of WDI time series (R4) by using SOM

Variable	R4C1S	R4C2S	R4C3S	R4C4S	R4C5S	R4C6S
TSS1	2	15	0	3	0	0
TSS2	0	4	7	2	6	1
TSS3	0	0	6	0	1	13
TSS4	2	10	6	0	2	0
TSS5	1	6	9	0	4	0
TSS6	0	6	6	6	2	0
TSS7	8	0	1	7	3	1
TSS8	6	12	0	0	0	2
TSS9	6	12	0	0	2	0
TSS10	11	0	1	4	2	2
TSS11	5	12	1	0	0	2
TSS12	11	0	1	4	2	2

Table 5.52 CS.8 clustering of WDI time series (R5) by using SOM

Variable	R5C1S	R5C2S	R5C3S	R5C4S	R5C5S	R5C6S
TSS1	0	3	0	0	0	0
TSS2	0	0	0	0	0	3
TSS3	0	0	2	0	1	0
TSS4	0	1	0	0	2	0
TSS5	0	3	0	0	0	0
TSS6	0	0	0	3	0	0
TSS7	0	1	2	0	0	0
TSS8	0	1	1	1	0	0
TSS9	1	1	0	0	0	1
TSS10	2	0	1	0	0	0
TSS11	0	1	1	1	0	0
TSS12	2	0	1	0	0	0

Table 5.53 CS.8 clustering of WDI time series (R6) by using SOM

Variable	R6C1S	R6C2S	R6C3S	R6C4S	R6C5S	R6C6S
TSS1	3	0	0	1	4	0
TSS2	1	1	1	1	4	0
TSS3	0	0	3	0	4	1
TSS4	4	2	2	0	0	0
TSS5	1	0	2	0	5	0
TSS6	1	1	4	0	2	0
TSS7	3	0	0	5	0	0
TSS8	3	2	0	2	1	0
TSS9	3	2	2	1	0	0
TSS10	4	0	0	0	4	0
TSS11	3	2	0	2	1	0
TSS12	4	0	0	0	4	0

Table 5.54 CS.8 clustering of WDI time series (R7) by using SOM

Variable	R7C1S	R7C2S	R7C3S	R7C4S	R7C5S	R7C6S
TSS1	10	2	16	0	3	17
TSS2	25	4	2	15	2	0
TSS3	2	3	3	21	19	0
TSS4	15	3	0	30	0	0
TSS5	19	0	3	25	1	0
TSS6	31	0	3	0	14	0
TSS7	39	0	4	5	0	0
TSS8	13	30	3	1	0	1
TSS9	11	34	1	0	2	0
TSS10	26	0	22	0	0	0
TSS11	12	30	4	1	0	1
TSS12	28	0	20	0	0	0

Table 5.55 CS.8 classification of WDI time series (Type-1 FIS)

FIS	Type-1											
Inputs	TSS1	TSS2	TSS3	TSS4	TSS5	TSS6	TSS7	TSS8	TSS9	TSS10	TSS11	TSS12
Outputs (subregion)	T1 FIS1	T1 FIS2	T1 FIS3	T1 FIS4	T1 FIS5	T1 FIS6	T1 FIS7	T1 FIS8	T1 FIS9	T1 FIS10	T1 FIS11	T1 FIS12
R1_R1	20	6	4	7	11	6	7	29	29	16	29	16
R1_R2	3	18	27	19	14	15	15	5	5	13	5	13
R1_R3	12	11	4	9	10	14	13	1	1	6	1	6
R2_R1	51	3	4	19	11	10	27	47	46	15	47	15
R2_R2	0	18	29	10	33	15	23	6	6	37	6	37
R2_R3	2	32	20	24	9	28	3	0	1	1	0	1
R3_R1	24	10	11	22	15	16	11	39	39	20	39	20
R3_R2	2	30	25	13	20	19	25	2	1	21	2	21
R3_R3	15	1	5	6	6	6	5	0	1	0	0	0
R4_R1	16	4	1	11	8	6	3	18	18	7	18	7
R4_R2	1	13	18	7	8	7	9	2	2	6	2	6
R4_R3	3	3	1	2	4	7	8	0	0	7	0	7
R5_R1	3	0	2	1	0	0	1	2	2	2	2	2
R5_R2	0	1	1	0	3	0	2	1	0	1	1	1

(continued)

Table 5.55 (continued)

FIS	Type-1												
R5_R3	0	2	0	2	0	3	0	0	1	0	0	0	0
R6_R1	2	2	1	3	2	2	1	5	5	4	5	5	4
R6_R2	1	5	4	5	1	4	4	2	2	2	2	2	1
R6_R3	5	1	3	0	5	2	3	1	1	2	1	1	3
R7_R1	10	26	4	11	19	31	30	46	45	17	17	46	17
R7_R2	19	11	25	24	4	14	15	2	3	31	31	2	31
R7_R3	19	11	19	13	25	3	3	0	0	0	0	0	0

Table 5.56 CS.8 classification of WDI time series (Type-1 FIS)

FIS	Type-1				
Inputs	T1 FIS1 T1 FIS7 T1 FIS8	T1 FIS2 T1 FIS3	T1 FIS4 T1 FIS5 T1 FIS6	T1 FIS9 T1 FIS11	T1 FIS10 T1 FIS12
Outputs (subregion)	T1_IR1	T1_IR2	T1_IR3	T1_IR4	T1_IR5
R1_R1	0	0	0	0	0
R1_R2	6	0	0	2	0
R1_R3	29	35	35	33	35
R2_R1	0	0	0	0	0
R2_R2	4	0	0	2	0
R2_R3	49	53	53	51	53
R3_R1	0	0	0	0	0
R3_R2	7	0	1	2	0
R3_R3	34	41	40	39	41
R4_R1	1	0	0	0	0
R4_R2	3	0	1	1	0
R4_R3	16	20	19	19	20
R5_R1	0	0	0	0	0
R5_R2	0	0	0	1	0
R5_R3	3	3	3	2	3
R6_R1	1	0	1	0	0
R6_R2	0	0	0	0	0
R6_R3	7	8	7	8	8
R7_R1	4	0	6	0	0
R7_R2	13	4	6	3	0
R7_R3	31	44	36	45	48

Table 5.57 CS.8 classification of WDI time series (Interval Type-2 FIS)

FIS	Interval type-2				
Inputs	T1 FIS1 T1 FIS7 T1 FIS8	T1 FIS2 T1 FIS3	T1 FIS4 T1 FIS5 T1 FIS6	T1 FIS9 T1 FIS11	T1 FIS10 T1 FIS12
Outputs (subregion)	IT2_IR1	IT2_IR2	IT2_IR3	IT2_IR4	IT2_IR5
R1_R1	0	0	0	0	0
R1_R2	21	31	6	11	21
R1_R3	14	4	29	24	14
R2_R1	0	0	0	0	0
R2_R2	49	34	11	13	41
R2_R3	4	19	42	40	12
R3_R1	0	0	1	0	0
R3_R2	25	37	16	9	27
R3_R3	16	4	24	32	14
R4_R1	1	0	1	0	0
R4_R2	12	19	8	8	10
R4_R3	7	1	11	12	10
R5_R1	0	0	0	0	0
R5_R2	2	1	0	2	1
R5_R3	1	2	3	1	2
R6_R1	1	0	1	0	0
R6_R2	0	6	1	5	4
R6_R3	7	2	6	3	4
R7_R1	7	0	12	0	0
R7_R2	16	38	1	18	37
R7_R3	25	10	35	30	11

Table 5.58 CS.8 classification of WDI time series (Type-2 FIS)

FIS	Interval type-2	Interval type-2	Interval type-2	Generalized type-2
Inputs	T1_IR1 T1_IR2	T1_IR3 T1_IR4 T1_IR5	IT2_R1 IT2_R2	IT2_R1 IT2_R2
Outputs (subregion)	IT2_R1	IT2_R2	IT2_RG	T2_RG
R1_R1	0	1	2	1
R1_R2	35	33	33	33
R1_R3	0	1	0	1
R2_R1	0	0	3	0
R2_R2	53	53	50	53
R2_R3	0	0	0	0
R3_R1	0	0	3	0
R3_R2	41	41	38	41
R3_R3	0	0	0	0
R4_R1	1	2	2	1
R4_R2	18	18	18	18
R4_R3	1	0	0	1
R5_R1	0	0	0	0
R5_R2	3	3	3	3
R5_R3	0	0	0	0
R6_R1	1	1	1	1
R6_R2	7	7	7	7
R6_R3	0	0	0	0
R7_R1	10	9	5	8
R7_R2	34	39	43	37
R7_R3	4	0	0	3

Table 5.59 CS.8 comparison of integration results between type-1 and type-2 FIS

FIS	Type-1 (T1)/interval type-2 (IT2)									
Inputs	T1 FIS1 T1 FIS7 T1 FIS8		T1 FIS2 T1 FIS3		T1 FIS4 T1 FIS5 T1 FIS6		T1 FIS9 T1 FIS11		T1 FIS10 T1 FIS12	
Outputs (regions)	T1_IR1	IT2_IR1	T1_IR2	IT2_IR2	T1_IR3	IT2_IR3	T1_IR4	IT2_IR4	T1_IR5	IT2_IR5
R1	6	9	0	0	7	15	0	0	0	0
R2	33	125	4	166	8	43	11	66	0	141
R3	169	74	204	42	193	150	197	142	208	67

Table 5.60 CS.8 comparison of integration results by using multiple Type-2 FIS

FIS	Interval type-2	Generalized type-2
Inputs	IT2_R1 IT2_R2	IT2_R1 IT2_R2
Outputs (regions)	IT2_RG	T2_RG
R1	16	12
R2	192	191
R3	0	5

Table 5.61 CS.8 prediction of WDI time series (NAR)

Variable	Average %RMSE	Best %RMSE	Worst %RMSE
TSS1	0.000027798	0.000013944	0.000048778
TSS2	0.000026712	0.000014912	0.000043324
TSS3	0.000064301	0.000038295	0.000111326
TSS4	0.000016197	0.000008287	0.000030772
TSS5	0.000012526	0.000006770	0.000020071
TSS6	0.000014930	0.000008680	0.000028881
TSS7	0.000134383	0.000084544	0.000201535
TSS8	0.000105706	0.000022670	0.000287229
TSS9	0.000091964	0.000025740	0.000286967
TSS10	0.000002059	0.000001169	0.000003839
TSS11	0.000110160	0.000024207	0.000252354
TSS12	0.000002194	0.000001190	0.000003500

Table 5.62 CS.8 prediction of WDI time series by GT2 region (NAR)

Variable	Average %RMSE	Best %RMSE	Worst %RMSE
TSS1	0.000420807	0.000092340	0.000949776
TSS2	0.000130195	0.000052815	0.000245523
TSS3	0.000332452	0.000167022	0.000591463
TSS4	0.000066602	0.000019978	0.000165994
TSS5	0.000137332	0.000070974	0.000296870
TSS6	0.000146327	0.000071138	0.000393558
TSS7	0.001108667	0.000573669	0.002632906
TSS8	0.000886138	0.000215622	0.002840459
TSS9	0.000851274	0.000180802	0.003587455
TSS10	0.000004522	0.000002657	0.000008226
TSS11	0.000960787	0.000170907	0.002616553
TSS12	0.000005042	0.000002598	0.000007630

Chapter 6
Discussion of Prediction Results with Neural Networks

Through the simulations carried out in the case studies, it is observed that both supervised neural networks NAR and NARX are viable methods for making predictions of multiple time series.

It was also demonstrated that it is possible to integrate results from different sources using Type-1 and Type-2 fuzzy systems, with which it is possible to establish similarities in the historical information of multiple variables between a sample of states or countries and combine global indicators.

We note that regarding the results of prediction of future values of the time series, we can mention that by using supervised neural networks it was possible to obtain the future values for each element or for a group of elements belonging to the same class.

For which we segmented the series temporal, and we observe that for certain variables better results are obtained when using the complete time series compared to using a data segment. The above depends on the period considered and the seasonal variation, trend, and cyclical variation, among others, of the time series.

Also, regarding do CS.1, one of the findings that is highly relevant, and we should highlight is that by combining intelligent methods, it is possible to focus on using an unsupervised model as the first phase to identify similarities or patterns in the data, seeking to highlight key variables and as a second phase focus on predicting the future values of these variables.

As well as focusing on the segments of an attribute for a given range, period, or geographic location, with which it is possible to plan strategies to improve transportation and roads in a given place.

We can also point out regarding CS.2 and CS.3, based on the tests carried out we can mention that it is possible to classify multiple sources of information using unsupervised neural networks, making it possible to consider qualitative and quantitative criteria.

P. Melin et al., *Clustering, Classification, and Time Series Prediction by Using Artificial Neural Networks*, SpringerBriefs in Computational Intelligence, https://doi.org/10.1007/978-3-031-71101-5_6

Considering the uncertainty implicit in the data, and subsequently perform the integration of the results using fuzzy systems, seeking to obtain a future measurement and evaluate the quality of the corresponding criterion or indicator, seeking to prevent aspects of respiratory health in the population.

Through the obtained results in the CS.4, as a first step, the aim is to separate and weight the obtained results by each supervised and unsupervised neural network model and highlight the relevant aspects that could support decision-making regarding the grouping and prediction of historical and future values.

Then, as a second step, use Type-1 and Type-2 fuzzy systems to integrate the results of neural networks and to classify the significant data selected by the user, seeking to obtain a general perspective for a place, entity, or an indicator simultaneously.

Now, for CS.5, we note that it is possible to predict the time series using multiple supervised neural networks based on partitions in the data formed based on the classes formed with the unsupervised neural networks.

Likewise, in the case study CS.6, it was observed that by using supervised neural networks it is possible to obtain clustering results of time series of countries and find similarities between historical data.

In addition, by using unsupervised neural networks it is possible to predict future values of time series and take advantage of the clustering results to make the prediction based on groups of countries that share similar information.

Based on the obtained results in CS.7, we note that because the data are ordered according to criteria that consider previously classifications, so, most countries presented very high similar results.

It means when using time series prepared based on the methodologies of international organizations, such as previously classifying each country, or considering certain criteria with based on the assigned, the fuzzy integration obtained similar results.

The aim is to develop a model to classify a sample of countries based on multiple criteria or indicators, using neural networks and hierarchical Type-1 and Type-2 fuzzy systems.

In relation to the CS.8, we note that through fuzzy systems we can integrate the obtained results through neural networks or create nested structures of fuzzy systems to do a level integration. Also, the Type-1 and Interval Type-2 fuzzy systems presented good results when considering two or more variables.

Furthermore, when integrating the results by Interval Type-2 and the Generalized Type-2 fuzzy systems, we observed that with the latter better results were obtained, by being able to separate a greater number of countries compared to the Interval Type-2 fuzzy system.

Chapter 7
Conclusions for Prediction with Neural Networks

In relation to the case studies addressed throughout this work, we can point out that the importance of time series analysis is essential to have objective information on how we manage resources, time, and make decisions.

Making good use of tools for managing indicators is a constant need in multiple international and national organizations, since by knowing how a country or state is performing in different areas, at the same time, it is possible to know in which areas it can improve.

For example, in areas such as education, health, institutions, environmental care, market functioning, infrastructure, among others.

Also, through the analysis of time series, it is possible to identify patterns or similarity in the information over time, as well as to predict future values, for which it is necessary to have technological tools that contribute to the timely integration of the results.

The evolution of some social and economic factors has contributed to the changes experienced by global components. Additionally, the constant changes in significant indicators represent new challenges and opportunities for organizations and governments.

Conclusively, in case studies CS.1 and CS.2, we observe that by using multiple Type-1 fuzzy systems it is possible to classify variables, elements, countries, states, or indicators.

It is also possible to integrate the obtained results using neural networks, which is intended to slightly simulate the behavior of cognitive functions in a person when they are deciding, always seeking to serve as a support tool in the decision-making process.

We also note in CS.3 and CS.4, that one of the advantages of the proposed method is that it is possible to combine artificial neural network models and use sets of fuzzy systems to perform classification, clustering, and time series prediction, by working or forming segments of information grouped by similar attributes.

© The Author(s), under exclusive license to Springer Nature Switzerland AG 2024   71
P. Melin et al., *Clustering, Classification, and Time Series Prediction by Using Artificial Neural Networks*, SpringerBriefs in Computational Intelligence,
https://doi.org/10.1007/978-3-031-71101-5_7

It allows us to obtain specific results for one or multiple variables. In addition, our model contemplates the management of uncertainty for decision-making and the integration of results through Type-1 and Interval Type-2 fuzzy inference systems.

Based on the obtained results in case studies CS.5 and CS.6, it has been shown that it is possible to use SOM and competitive neural networks to find groups based on the similarity of historical data trends between countries.

In addition, it is possible to use NARX and NAR supervised neural networks to perform time series prediction: population, urban population, particulate matter (PM2.5), carbon dioxide (CO_2) and Covid-19 cases.

We can conclude that in case studies CS.7 and CS.8 the simulation results demonstrate that it is possible to use Generalized Type-2 fuzzy systems to integrate the indicators of multiple countries in a better way than with Type-1 and Interval Type-2 fuzzy systems.

As a result of global uncertainty, the interest shown by nations to promote sustainable development, care for the environment and improve people's well-being.

It is essential to increase efforts to promote programs aligned with goals and objectives that consider integral aspects where innovation plays a fundamental role, being the study of numerical and non-numerical historical data, to achieve a broad-spectrum qualitative and quantitative integration.

We propose as future work consider selecting global datasets consisting of well-being indicators, to perform tests of multiple fuzzy integration techniques.

We also contemplate developing new case studies that consider optimization methods applied to fuzzy systems, particularly to optimize fuzzy rules and membership function parameters. It is also intended to improve prediction results by using intelligent hybrid techniques to perform classification and prediction tests.

Index

© The Editor(s) (if applicable) and The Author(s), under exclusive license
to Springer Nature Switzerland AG 2024
P. Melin et al., *Clustering, Classification, and Time Series Prediction by Using Artificial
Neural Networks*, SpringerBriefs in Computational Intelligence,
https://doi.org/10.1007/978-3-031-71101-5